Tone: A Study in Musical Acoustics

Siegmund Levarie and Ernst Levy

TONE

A Study in Musical Acoustics

Second Edition

Published by The Kent State University Press

Copyright © 1968, 1980 by Siegmund Levarie and Ernst Levy.
All rights reserved
Published by The Kent State University Press, Kent, Ohio 44242
Library of Congress Catalog Card Number 80-16794
ISBN 0-87338-250-1
Manufactured in the United States of America

Library of Congress Cataloging in Publication Data

Levarie, Siegmund, 1914-
  Tone.

  Bibliography: p.
  Includes index.
  1. Sound. 2. Music—Acoustics and physics.
I. Levy, Ernst, 1895-   joint author. II. Title.
ML3807.L54   1980   781'.22   80-16794
ISBN 0-87338-250-1 (pbk.)

To Hugo Kauder

TONE, from Greek *tonos*: rope, cord, tension, stretching, exertion of force.

Preface

This book serves primarily the purpose of investigating tone in terms of musical values. Against the tendency to demonstrate "the science of music," a subtitle like "the art of acoustics" might well have been chosen. Whatever the scientific involvements of the contents, the authors think of themselves as musicians writing for musicians and music lovers.

The implications and risks of the two opposite approaches may be illustrated by a famous controversy in the field of optics. Newton explained the phenomenon of color as a function of the mechanics of light rays. According to him, color exists objectively in the nature of light. This one-sided causality has a great scientific appeal but leaves the artist unsatisfied. Goethe obviously did not deny the possibility of a mathematical explanation of light and color; but contrary to most physicists since Newton, he placed the main emphasis on a comprehensive physiological and psychological interpretation of light. According to him, color is created by the interplay of light and darkness, of light and not-light, of positive and negative. To Goethe (and after him to philosophers like Schelling and Schopenhauer), color is a manifestation of polarity.

The situation is similar to that in the realm of acoustics. Here the scientific attitude of the last two centuries has been generally accepted as supplying a total explanation of the phenomenon of tone, whereas it supplies at best one-half of it. Our book follows Goethe in more than just general attitude. A short document by him exists, part of his correspondence with the Berlin *Kapellmeister* Karl Zelter, that outlines a theory of acoustics. In contrast to the rich and voluminous data presented by Goethe's studies on optics, his papers on acoustics are sketchy and brief. But the essentials are clearly stated. The orientation, as in his theory of light, is gained from the concept of polarity. Tone is postulated as a given inner phenomenon. Hence, the main question is not what the phenomenon is but how it comes into being. After outlining the premises, Goethe presents the subject matter in three parts: organic (ear, voice, and body rhythm), mathematic (monochord), and mechanic (instruments). All sections point toward a contemplation of the work of art. In general conviction as well as in some specific details of organization, this book follows Goethe as a guide.

The authors are more immediately indebted to the contemporary Swiss philosopher Hans Kayser (who died in 1964 while this manuscript was being prepared). In his extensive writings, Kayser endeavored to develop an epistemological tool by proceeding from acoustics. He deserves credit for the modern exposition of the Pythagorean table (which had been rediscovered by Albert Freiherr von Thimus around the middle of the last century); its successful application as a key to Pythagorean philosophy and to number symbolism; an explanation of the harmonical method used by Kepler in establishing his astronomic laws; and, in general, the uncovering of a harmonical background in a wealth of phenomena.[1] The connection with Goethe is evident from Kayser's *Harmonia Plantarum* (1943), for instance, where the norms recognized in and through tone are applied to the vegetable kingdom. But above all, Kayser—first against the modern scientific tendency to overemphasize the quantitative aspect of the world—insisted upon the autonomy and reality of tone as a value and upon the consequent "musicalization" of number. Without Kayser, the teaching of acoustics would probably still be what it was when the authors of this book were music students: a peripheral course dealing with some elementary notions about the physics of sound, lacking in inner significance because unfit to bridge the gap between physics and music, and conveying little importance to the musician.

In many ways, this book is an outgrowth of a course taught by the authors and some of their colleagues at Brooklyn College over the last ten years. We consider the course a unifying experience which supplies the students with principles rather than techniques, with insights rather than rules—in short, with the fundamental facts to which all further musical activities can be

---

[1] Throughout this book, the term "harmonic," which belongs to music theory, is kept distinct from "harmonical," which has wider connotations that will become gradually more definite as the reader proceeds.

related. We think of acoustics as being highly practical and formative for students.

For the present purpose, we have no intention, unless necessitated by the context, of duplicating standard information available in any good physics textbook. This does not mean that the book at hand is not complete in its own way. But for a fuller discussion of much relevant material, the student should turn to appropriate works in the field. Suggested readings are listed at the end of this book.

Beyond the immediate musical lessons, acoustics leads us to thoughts that have a bearing on our total life.

First, the study of acoustics demonstrates the advantages and virtues of returning to fundamentals. The monochord has *one* string, and the experiments on it represent *one* kind of operation. There is concentration on the basic phenomenon of a single tone in relation to other tones. The complications that arise are endless; but all music, complex as it can become, is covered by the definition just given: the relation of one tone to other tones. The immediate future of music and perhaps of our life depends, we submit, not on a more refined splitting of the octave, or on a more complex electronic machine, or on more ingenious experiments in all directions, but on a return to fundamentals.

Second, acoustics teaches the need for norms—better still, the recognition that norms exist whether we like them or not. The octave is a norm. One of our colleagues, Robert Sanders, once said to a class: "You can't argue with an octave." The triad is a norm. There are others that emerge strongly and logically as one studies acoustics. Music cannot exist without norms. A sensible course of action is not to fight the norms but to find out how best to operate in relation to them. This is the real meaning of harmony. Freedom, as we learn sooner or later, is never absolute but rather the result of the acceptance of norms—be they norms of counterpoint in music or of behavior in ethics. For what is the alternative to norms? Not freedom, and not art, but only nihilism and chaos.

Third, a universal lesson to be drawn from acoustics is the significance of hierarchy. We mean a hierarchy of values and not of statistics—a hierarchy of qualities (in which good art is rated above less good art) and not merely of

x | Tone: A Study in Musical Acoustics

quantities (in which eight cylinders are automatically deemed preferable to six). Tones, we learn in acoustics, are produced in a definite order on the vibrating string. The octave always comes before the fifth, the third always before the seventh. Intervals are not equivalent. We hear that they are different, and the terms consonance and dissonance indicate the reality of the value scale. Apparently modern man who loves statistics is afraid to believe in a rank order of values. Living in a democracy, he might do well to remember that the ideal interpretation of this concept certainly does not consist in a leveling of values, of taste, and of standards of behavior. Rather, it stems from the recognition, so demonstrable in acoustics, that everything has its definite place in relation to everything else and, by finding it, reaches its full individual and social potential.

Brooklyn College
of the
City University of New York
November 1967

Preface to Second Edition

A second edition of a book offers authors, apart from the inherent satisfaction, a chance to correct and revise. The "Suggested Readings" list (pp. 250 f.) has been slightly expanded. The two Appendices (pp. 241 ff.), practically tested in many seminars since the appearance of the first edition of this book, are new. Throughout the text, small errors of various kinds have been silently set right.

The discussion of the term-pair "static-dynamic" (pp. 209 f.) requires a modification gained from the authors' experiences and thoughts during the past decade. A distinction seems necessary between inherent qualities and induced qualities. For the first, the concepts "static-dynamic" suffice. To mark the antinomy within the latter, we suggest the terms "ontic" and "gignetic" (from the Greek roots for, respectively, "being" and "becoming"). The following quotation explains: "In an ontic position, we abstract from time; we consider things *sub specie aeternitatis*, under their typical permanent aspect, not subject to change. In a gignetic position, on the contrary, we consider phenomena according to their temporal nature, changing and fugitive. When we speak of a baby, of an old man, a river, a musical scale, a chord, we speak of the being of things. When we speak of development, of getting old, of erosion, of a scale rising or falling, of a chord progression, we speak of the becoming of things. When you swim fighting against the waves, you experience gignetism. Rise high enough to perceive the contours of the lake or the sea, and you will discover that the movement of the water was, after all, no more than an agitation within an immobility. Thus one will have experienced successively the gignetism and the ontism of the same phenomenon.

"Now these two modes of apprehending phenomena are reflected in the things themselves according to whether their nature participates more or less in the one or in the other. Music offers revealing examples. A consonance is by its nature ontic whereas a dissonance is naturally gignetic. Yet if one employs a perfect major triad as a dominant, one thereby confers a gignetic aspect to a consonance. Inversely, by ending a piece on a dissonant chord, one thereby confers on it an ontic aspect."[1]

[1] Ernst Levy, "Aperçu sur un arrière-plan de l'histoire de la musique," *Revue musicale de Suisse Romande* 31 (1978), 72-92. Trans. S. L.

Acknowledgments

A campus becomes an academic community in the best sense when various members center their attention on the work of a colleague (let alone of two colleagues). Generous encouragement and severe criticism are dispensed with equal grace, competence, and—most important—patience. In this spirit, Professors Robert L. Sanders and Ernest G. McClain, of music, and Professor Edward H. Green, of physics, read the first draft of the manuscript cover to cover. While thus demonstrating charity toward the authors, they displayed no clemency toward the text. If the hours of discussion that followed seemed endless to them, we wish that our gratitude might convey a similar impression.

Professor Josef Mertin, of the Akademie für Musik und Darstellende Kunst in Wien, Austria, shared his enormous knowledge of instruments with us whenever we turned to him for clarification of a special point. Ernst Schiess, of Bern, Switzerland, offered unique assistance for the section on bells. Jean Hakes measured our theoretical knowledge against her practical experience as a singer.

Beyond all, we know that our work progressed accompanied throughout by the beneficial spirit and wisdom of Hugo Kauder. The dedication of this book is only a small sign of our devotion.

Credits

The monochord (Fig. 8) was built by Professor Ernest G. McClain.

The Physics Department at Brooklyn College, under the initiative of Professor Edward H. Green, produced original photographs of sine and noise curves (Figs. 2 and 3).

The Journal of the Acoustical Society of America made available the photographs of the piano sound curves (Fig. 43).

Professor Harold C. Jensen, of Lake Forest College, furnished new Chladni figures, which were photographed by Milton Merner (Fig. 59).

Ernst Schiess gave us free access to his large private collection of drawings and photographs of bells. The illustrations concerning bells are all his generous contribution.

What we need is not to "know" the truth but to experience it.
                —C. G. Jung, *Seelenprobleme der Gegenwart* (Zürich, 1931)

Contents

PITCH DESIGNATIONS

Throughout this book, octave ranges are named according to the following system:

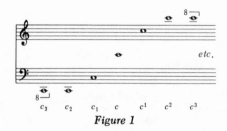

Figure 1

Capital letters (C, D, E, etc.) are reserved for general pitch names without regard to a specific octave range.

1 | Preamble

PREMISES

Music is not, as some acousticians would have us believe, "something that happens in the air." It is something that, first and last, happens in the soul. To an outer physical something corresponds an inner spiritual something: tone. Music happens when both are "attuned" to each other.

Tone, as are all sensations, is a psycho-physical fact. Therefore no intensity of intelligence, no amount of imagination, can be a substitute for the experience of music. In a theoretical investigation of the raw material of music, this experience assumes the form of experiment. In this sense, the method of our investigation will be prevailingly inductive. Although this method might be called "scientific," as the term is used in the natural sciences, the rating of the results in music differs radically from the rating of the results in science. In physics, the rating bears on quantities. Hence the findings are universally acknowledged. In music, the rating bears on qualities. Hence the investigators run the risk of being subjective. A musician, whether theoretical or practical, must quickly face the fact that he cannot eschew evaluation. The substitution of quantities for qualities would be an absurd and futile endeavor. Subjectivity and its risks are inherent in music, as they are in all matters humanistic.

Because tone is a psycho-physical fact, we obtain musical insights on the psychological level by investigating the natural structure on the physical level. The data on both levels will then become mutually symbolic. In basic physical experiments, we treat tone as a natural phenomenon and, appropriately to any object of science, express the findings in mathematical terms. These mathematical data then become symbols of musical facts and, in turn, acquire a meaning of value. A number thus charged with a qualitative meaning may be called a "musical number." The process of this kind of investigation eventuates in the establishment of two consistent, parallel, and mutually symbolic systems.

From this viewpoint, the term "symbolic" should be taken quite literally without any esoteric connotation. The Greek verb *symballein* means a 'throwing together' of several things—not a poetic representation of one by the other, nor a causal explanation. The musical quality of a tone and the mathematical data concerning the same tone are symbolic of each other: they happen to coincide but they do not act on each other as cause and effect. In the words of Victor Goldschmidt: "Our capacity to apprehend the outside world may be explained thus: that there are processes in our mind (microcosm) which are analogous to those in nature (macrocosm). These psychological processes we call natural laws."[1]

SOUND AND VIBRATION

In this book we intend to deal with musical phenomena and address ourselves to musicians. The title holds forth this promise by restricting itself to tone instead of embracing the whole concept of acoustics. What is the difference between these two terms? Nothing less than that between music and physics, between art and nature, and between limitation and infinity. To reach an understanding of tone, we must first define the acoustical phenomena on which tone depends and from which it derives.

Acoustics is the science of sound. The Greek root *akouein* means 'to hear.' A theory of sound develops laws for all that is heard. Sound is vibrational energy of the kind that can be apperceived and transformed by the ear. This is a qualitative definition, bound by the subjective sensation of anybody's ear. Obviously, there are other kinds of to-and-fro motion that cannot be heard. If we push a swing back and forth, we set off an oscillation that we do not hear. If we turn on a light, we set off a vibration that we do not hear. The swing and the light have each disturbed an equilibrium by a pulsating motion which is apperceived by some senses other than the ear. We can feel the swing if it hits us, and we can see the light if it reaches our eyes. In order to be

---

[1] *Ueber Harmonie und Complication* (Berlin, 1901), p. 1. Transl. by the authors.

heard, the vibration must meet certain limiting conditions. The swing has not moved to and fro often enough in any given time unit; the light, on the other hand, too often, to appeal to the ear. In anybody's experience, wind may increase in speed from a gentle breeze that is felt to a "howling" gale that is heard.

Among an infinite possibility of physical vibration, sound is a vibration within a certain range. The number of vibrations per second is called "frequency."[2] The frequencies heard by the human ear lie approximately between 16 and 20,000. The figures vary with the individual and even with the same person's age and health. A middle-aged person seldom hears frequencies above 15,000. An older person is likely to have a still lower ceiling. In short, this mathematical definition is not much more precise than the qualitative definition with which we started.

Just as we have gained the concept of sound from the infinity of possible vibrations by letting the ear set physiological limits, so we may now distill the concept of tone from the wider one of sound by letting the ear set musical limits. This is done by eliminating those sounds that are musically of no primary significance, such as noise and language. Both, of course, may be used in the context of a musical composition, but they are not of the essence.

Tone is that part of sound that we call "musical." In terms of frequency, one already detects a further setting of limits; for the highest pitch in the orchestra, c^4 on a piccolo flute, has a frequency of only 4224. But there is a more significant qualitative distinction between tone and noise. Both are sound and hence the result of a vibration; but the vibration of tone is regular, whereas the vibration of noise is irregular. Tone is an orderly phenomenon, noise a disorderly one. Thus tone is a distillate from noise.

A regular oscillation is best illustrated by a swinging pendulum. It moves in perfect periodicity. Physicists, as much as musicians, have shown their sensitivity to the special quality of tone by naming the regular vibration producing it a "simple harmonic motion." The vibration can be recorded for the eye by a pencil attached to the bottom of the pendulum and writing on paper

---

[2] Generally abbreviated as *cps* (cycles per second) or *Hz* (Hertz, in honor of the discoverer of radio waves, Heinrich Hertz).

that moves at right angles to the direction of the vibration. The result is a regular and symmetrical curve, known as a "sine curve":

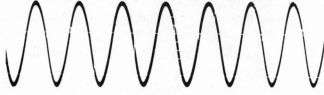

Figure 2

Noise thus recorded would produce an erratic and irregular graph:

Figure 3

One more step in the continuing process of limitation is necessary to give the musician the material with which he can work. He has to select the tones he wants to use. The strong emphasis on the necessity of limitation, which has guided these considerations thus far, reflects not a subjective prejudice but a fundamental artistic law. There is no art without limitation. The sculptor working in marble has set his limit by the choice of this material to the exclusion of all other materials. The playwright putting an English king on the stage limits himself—if he is a good playwright—to those episodes that may form an artistic whole. The point, amply discussed in aesthetics, need not be further labored here. Goethe sums it up:

> *Das ist die Eigenschaft der Dinge:*
> *Natürlichem genügt das Weltall kaum;*
> *Was künstlich ist, verlangt geschlossnen Raum.*

This is the property of things: the All
Scarcely suffices for the natural;
The artificial needs a bounded space.
(*Faust* II. 6882-6884)

Let us proceed from any vibrating body producing a tone, e.g., a string we pluck. We accept the body (in this case, the string) as a given entity, and we call the tone that this entity produces "the fundamental." We do not take long to discover that a shortened string raises the pitch. In general terms, physical reduction of the given entity renders a higher tone. Inversely, we can conclude that a physical enlargement of the given entity renders a lower tone. The shape of a grand piano illustrates this point concretely; and most children know that a tapped glass sounds lower as one pours water into it, that is, as one enlarges the vibrating mass.

By shortening the string gradually, we do not get clear relations between pitches. The glissando produced by a continuous shortening of the string is a pitch motion, a continuous "becoming." The number of pitches distinct from the fundamental lies in infinity—the realm in which artistic creation is impossible. The nature of a glissando is of a special order. During the Second World War, the Germans are said to have used a siren-like device on their dive bombers, the howling of which was meant to intensify the horror of the attack. Today the howling of sirens assails us from all sides: fire engines, police cars, ambulances, air-raid drills. Has the intended shock worn off so that we accept the wailing factory siren as a substitute for the noon bells of a church? Our healthy and elementary reactions to acoustical events are vanishing; and with them goes our feeling for acoustical symbols that lie at the psycho-physiological root of music. The glissando is an acoustical symbol for dynamism run wild; and if our spontaneous reaction to the howling of the siren is one of fright, the reason lies in the demonic character of the glissando. Some feeling for this basic truth must have subconsciously prevailed when the decision was reached to identify the alarm signal with a wailing sound, and the all-clear signal with a steady tone.

As we turn from the total glissando that traverses the whole audible range in continuous pitch variation to the limited glissando that occurs between two fixed tones, we notice a remarkable change of the effect. The horror has vanished. In its place, we gain an impression of indolence, of a somewhat nauseating sweetishness, and of a shock based on contrasts. A singer's or violinists's bad portamento falls in this category. What accounts for this kind of impression? First, against the fixedness of tones, the glissando represents an element of naturalism. The fall from a higher to a lower level is experienced as a shock, similar in character, for instance, to the feeling when one passes from the sung word to the spoken word. Second, the glissando represents an element also of the irrational in the midst of the beautifully ordered. Hence the portamento appears both somewhat ridiculous and sentimental, comparable to the tragicomic but always painful spectacle of a performance by tamed lions.

Clearly, the glissando has a place in music only as an exceptional effect. As such, it can be used advantageously. But the slightest lack of sureness in taste will be fatal, the danger growing in direct proportion to the width of the interval bridged by the glissando.

The alternative to shortening the string gradually and continuously is shortening it with the help of a guiding principle which will establish relations between individual tones. If we have called the glissando a "continuous becoming," a musically undefined state, we can call the tone a "being," a particular musical element. Relations between tones, as between defined individuals, are characteristic and fixed. At this point one need not ask for the exact finite selection of artistically usable tones from the infinity of the physically possible tones in the glissando. The passing of centuries and civilizations teaches us that different selections are possible. But without a principle of selection, without the acceptance of a limitation, the world of tones would remain a concern of physics and not of music.

The following diagram attempts to summarize how, from the infinite world of physical vibration, one reaches the finite world of musical tones by a process of repeated distillation. The left column shows the increasing purification; the right column, the nonmusical elements that are eliminated at each step:

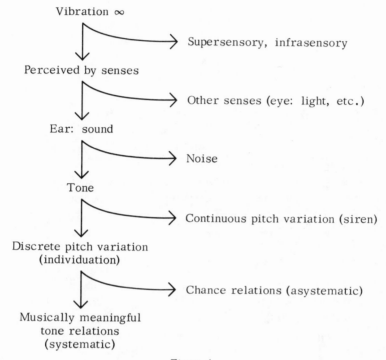

Vibration ∞

→ Supersensory, infrasensory

Perceived by senses

→ Other senses (eye: light, etc.)

Ear: sound

→ Noise

Tone

→ Continuous pitch variation (siren)

Discrete pitch variation
(individuation)

→ Chance relations (asystematic)

Musically meaningful
tone relations
(systematic)

Figure 4

The transition from physics to music, from nature to art, from the observation of the continuous pitch variation in a glissando to the discovery of the discrete pitch in the individual tone, is not a gradual process leading from the becoming to the being. On the contrary—everywhere we first perceive and deal with separate units, fixed individuals, discrete quantities. Mathematically speaking, the whole numbers were discovered or invented long before the "infinitely small" began to be apprehended. The distinctness of the relation between tones will provide us with a guiding principle in our search for an acoustical understanding of musical norms.

Exercises

1. Check the upper and lower limits of your hearing. You will find the apparatus that does it easily and quickly in the laboratory of a physics department, or the office of an ear specialist, or the display rooms of a telephone company.

2. Go back when you have a cold and compare the results.

3. Discuss the applicable principle of limitation with reference to concrete examples from all the arts. What limiting, that is, formal, elements set art apart from life? What is the difference, for example, between the historic Henry IV and Shakespeare's? Between a person and a sculpture of that person? Between a landscape and a painting of the same landscape? What is the function of the frame around a painting? What limiting principles can you isolate in a Beethoven symphony? What distinguishes an opera from the novel on which it is based? Et cetera.

4. Construct a sine curve. To do so intelligently rather than merely mechanically, remember first that the motion of a pendulum, or simple harmonic motion, is describable in terms of direction, and of rates of velocity change. Even a casual observation of a swinging pendulum discloses that at the point of maximum amplitude, where the reversal in direction takes place, there is no motion for a moment, whereas at the point where the pendulum passes the vertical position the motion is swiftest. In Figure 5 below, a point P is rotating at constant speed in a circle with the radius r.

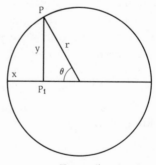

Figure 5

Point P_1, which depends on P, wanders on the horizontal axis x with variable speed. The speed is greatest at the moment when the angle θ at the center is 90°, and then decreases toward zero at $\theta = 180°$. The process is repeated until θ returns to the initial position of 360° = 0°. The point P_1 is performing a simple harmonic motion. You may think of it as a piston in a steam engine or in an automobile motor. Presently consider y and r. The radius r does not change. Let us call it "1." The length of y, however, changes as a function of θ. As θ increases from 0° to 90°, y increases from 0 to the length of r, that is, 1. The rate of increase is not uniform, as we have seen, but all values of y for $\theta = 0°$ to $\theta = 90°$ will lie between $y/1 = 0$ and $y/1 = 1$. The relationship of y/r, which is independent of the size of the circle, is called the "sine" of the angle θ. Simple harmonic motion is accordingly defined as a motion of which the rate of change is measured by that of the sine of an angle when the angle is increasing at a uniform rate.

A graph of a sine curve can be constructed by letting time be represented in it. Figure 6 shows how it is done.

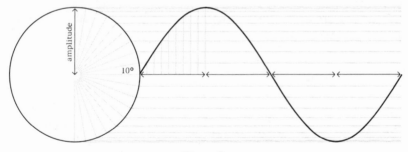

Figure 6

Draw a circle of any radius. The length of the radius represents the amplitude of the vibration. With a protractor, divide the right half of the circle into equal parts. The finer the division, the more precise the drawing of the curve will be. Draw parallels to the horizontal x-axis from each division point on the circumference. Choose any horizontal distance as a time unit, record on the extended x-axis at least twice as many units as there are divisions of the half-circle, and draw vertical parallels through the time points. Drawing, by the way, is greatly simplified by the use of graph paper. Now plot the graph as shown. Mark the intersections of the horizontal and vertical lines, beginning at that of the x-axis and the circumference, and proceeding to right for the time units and up and down according to the parallel circle divisions. Draw the curve by connecting these points.

5. Construct three sine curves in the proportion 2: 3: 7. Plot and draw the resulting curve. The combination of simple sine waves forms complex waves. The analysis of a complex wave, as developed by the French mathematician J. B. J. Fourier around 1800, is a complicated operation; but the inverse process of combining graphs of different sine curves, is not. Figure 7 on the next page shows two waves in the frequency ratio 1:3, and the resulting complex wave form. It is produced by the addition of the positive and negative distances from the x-axis (i.e., the distances above and below at any one point, or simply of the values of y). Make this addition for a sufficient number of points to plot a smooth curve.

6. Construct three sine curves in the proportion 2: 5: 9. Plot and draw the resulting curve.

7. Construct three sine curves in the proportion 3: 4: 10. Plot and draw the resulting curve.

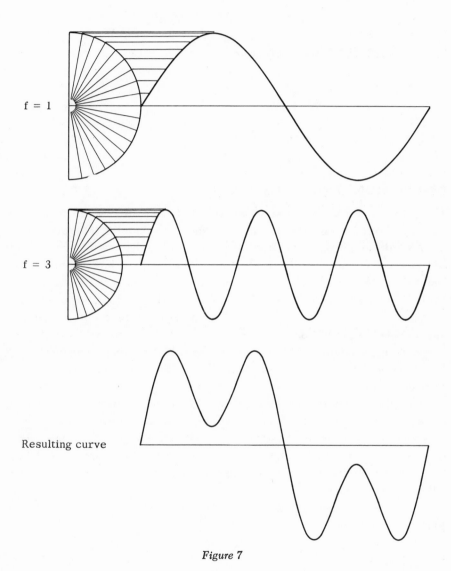

Figure 7

2 | The Monochord

DESCRIPTION AND CONSTRUCTION

There are several ways by which an acoustician might proceed to investigate the nature of tone and of tonal relations. For almost 3000 years now, a monochord has proven its usefulness for musical and scientific experiments and demonstrations. The Greek word *monochord* means, literally translated, 'one string.' The string is usually stretched across a soundboard in order to be easily heard and also measured. Although one string will do, several strings permit us to retain one string as a control and to hear the results of complex measurements simultaneously. To avoid a semantic confusion, the term "sonometer" might be substituted for a monochord that has more than one string. Retention of the traditional term "monochord" is justified, however, because all strings involved are tuned to the same tone.

Although not only the number of strings but also the length of the string is arbitrary, two practical considerations enter. A total of thirteen strings permits the setting of any twelve-tone scale including the return to the starting point at the octave; and a length of 120 units (preferably centimeters) facilitates the demonstration of relationships involving at least the first six integers (all of which are divisors of 120).

A three-string sonometer, adequate for most initial experiments, can be constructed at home.

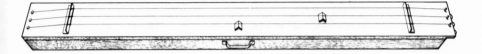

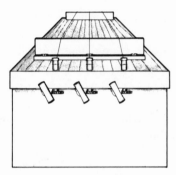

Figure 8

On one end of a board 1" × 6" × 5', mount a three-screw guitar machine (recessed about 3/16" to permit its posts to protrude sufficiently). Anchor the strings to screws near the far end; number 6 piano wire (.016" diameter) is recommended.[1] Cut two bridges, each 4½" long, of angle aluminum, 3/4" × 3/4" × 1/8", and center them on the board facing each other exactly 120 cm apart. Movable bridges for dividing the strings are easily made from 1" lengths of the same material if 3/4" wood corner molding is glued to the inside so that the apex of the angle is slightly higher above the board than the strings. A frame 3" wide glued around the underside of the table will greatly strengthen it and discourage warping.

For practical reasons, the strings should be tuned to c_1, one octave below middle c, in order to make divisions of the string by high numbers demon-

---

[1] Available in quarter-pound coils for about one dollar from the American Piano Supply Company, Main Avenue and South Parkway, Clifton, New Jersey.

strable by a string portion not too short for comfort. The various divisions of the string that our studies will require should be recorded, together with the musical results, on strips of millimetric graph paper 120 cm long. These strips can be easily fitted under the strings, fastened with masking tape, and readily exchanged and stored.

DIVISION OF THE STRING

The discreteness of pitch postulated earlier (cf. pp. 6 f.) parallels the application of whole numbers to the shortening and lengthening of the vibrating string. Integers are not, as some people claim, merely an ultimate abstraction from matter. They are rather an invention applied to nature. They are the fundamental principle of organization which man brings to bear on the whole cosmos.

We begin by dividing the string by integers. As a convenience, the pitch of our string is assumed to be C, although absolute pitch is not of the essence.

DIVISION BY 2

We stop the string exactly in the middle. The eye alone will at best approximate the dividing point. A measuring tape is needed. There is no foretelling what pitch the halved string will produce. We pluck either side, compare the new pitch with that of the full length, hear that the new pitch is higher, and accept the relation of the two pitches as the interval known as the octave.

The term "octave" at this moment is very misleading. There is an intrusion of the number 8 although the string has been divided by 2. The confusion is unfortunate. It is the result of mixing an absolute principle—in this case, the division by 2—with an arbitrary convention—in this case, the filling of the gained tonal space, including the boundaries, by eight tones in Western tradition. The Greeks were more sensitive to this phenomenon and happily avoided the confusion by calling the interval formed by the whole and the halved string the "diapason." This word, literally translated, means 'through all,' 'through the whole'—through the entirety of the tonal space, we would say. All basic tonal relations can be found within the diapason, within the "entirety" of the octave. Beyond it, they repeat themselves. This fact is recognized not only in

Western tradition but in all civilizations of all times. The interval C to E for instance, heard as a tenth presents a modified appearance of a major third but certainly not a radically new experience. The two tones forming an octave achieve an identity that transcends their distinctness.

The literal meaning of the term "diapason," evoking a wholeness, indicates the supreme significance of this interval. Our division of the diapason into eight steps (i.e., seven different tones), into an "octave," accustomed as we are to it, is arbitrary. It is doubtless related to old cosmic convictions: the division of the year in the ancient Near East into 7 seasons of 7 weeks each, containing 7 times 7 holidays; the number of days in our week; the number of planets in the solar system; et cetera. Other civilizations are oriented toward other numbers, that is, toward other principles of organization. The pentatonic scale of the Chinese is familiar even to Westerners; and the five tones within the diapason (how nonsensical to speak of "the five tones within the octave"!) are easily related to the preponderance of the number 5 in Chinese mythology and wisdom: 5 founders of the realm, 5 points of heaven, 5 elements, 5 colors, et cetera. Other civilizations favor still other organizing principles which are reflected in the number of tones employed within the diapason: the Indian scale contains 22 steps; the Arabic, 17.

The diapason, gained by the division of the string in 2, represents a cosmic absolute. All music systems are based on it. The number of steps within the diapason, on the other hand, reflects a man-made conditioning. Musical languages vary accordingly.

The indebtedness of the diapason to the number 2 raises one more question before we can give our full attention to the musical interval gained by the initial division of the string. What is the relationship of "2" to "the whole"? Does the number 2 have special privileges denied to other numbers? Yes, it does. Just as the division of the string in 2 was necessary to create the basic tonal space, so the division of the universe into duality was necessary to create the world. The Biblical story of the Creation is one of repeated division: heaven and earth, light and darkness, land and water, male and female. Duality or polarity pervades our life. It makes life possible and it defines life. It determines the framework within which all problems arise and within which they demand a resolution.

Let us return to our string divided in 2. The new pitch sounds against the old pitch the interval of the octave. The half string forms against the whole string the measured relation of 1:2. Hence we can state that the musical quality of the octave is bound to the mathematical quantity of 1:2, or, octave $= 1/2$.

The practical applications of this law are immediately obvious to the musician. A violinist wishing to play the octave of an open string will put his finger exactly in the middle of the string. An organ builder finishing a pipe to sound an octave above another similar pipe will cut the higher pipe one-half the length of the lower. A child producing a tone by blowing into a bottle will find the octave by filling half the bottle with a liquid.

The violinist rather than the pianist will be aware of another practical consequence of the law which our first experiment has established. If the octave corresponds to the relationship of 1:2, then the same interval will sound whether the absolute measurements involved are 1 and 2, or 10 and 20, or 20 and 40, or 100 and 200. The relationship of each of these pairs remains constant although the difference changes. Each relationship above represents an octave whether the difference in string length be 1 (2 minus 1), 10 (20 minus 10), 20 (40 minus 20), 100 (200 minus 100). An interval, in short, is a relation of two tones rather than the distance between them. The Greeks, again, were more sensitive in this respect than we are; for they used the term *diastema* to keep the concept of distance distinct from that of the basic intervallic relationship. Today we have come to apply the term "interval" indistinctly to both relationship and distance; and this fuzzy practice may be either the cause or the result of much confused thinking about intervals.

A violinist, in any case, knows that on his string he must finger ever smaller distances as he plays ever higher octaves. Assuming that the vibrating length of his A-string is 13 inches, he will play the octave above this open *a* 6½ inches from the end; the octave above that open a^1, 3¼ inches higher up; the next octave, 1⅝ inches higher up; and so forth.

A pianist is likely to be startled by this physical truth, for to him the distance between the two tones of an octave remains the same regardless of where he plays the octave. The explanation lies in the mechanism of the piano. A keyboard, by its particular construction, transforms the relationship into a distance. It transforms the division into subtraction, and multiplication into addition. Thus it acts like a table of logarithms. This simplification of procedure is reflected also in our system of notation where every interval, regardless of absolute pitch, is always shown by the same distance on paper. The fact is that we hear in two different ways: melodically and harmonically. Two tones heard melodically constitute a distance, but the same two tones heard harmonically constitute a ratio. The melodic distinction between the two phrases below is gigantic; the harmonic distinction, as far as the ratio of the two tones is concerned, minimal.

Figure 9

If division of the string by 2 produces a pitch one octave higher, we may reason that multiplication of the string by 2 produces a pitch one octave lower. By taking the halved string as a starting point, we can hear the lower octave on the monochord by sounding the whole string. Multiplication of a given string is, of course, a practical impossibility, for a string cannot be made longer than it is. The thought, however, is tempting; for it is generated by the recognition of a spiritual principle. Division and multiplication are reciprocal operations. To every up corresponds a down. The tones gained by multiplication are a psychological reality, whatever the physical limitations.

The life force of the 2nd partial, that is, of the tones gained, respectively, by division and multiplication of the string by 2, may be represented by the following diagram:

Figure 10

We have set up the equation: octave = 1/2. Read this way, the equation satisfies many people who consider the concept of 1/2 a complete definition of the octave. Yet, this definition, though correct, is painfully defective. An intelligent man, grasping all we have said thus far, will have at best an imperfect notion of the octave if he has never heard one—if he had been born deaf. Such was the fate of Joseph Sauveur (1653-1716), who yet gave the first scientific explanation of overtones and the first calculation of absolute frequencies. Whitehead states the point succinctly, persuasively, and beautifully: "What is wanted is an appreciation of the infinite variety of vivid values achieved by an organism in its proper environment. When you understand all about the sun and all about the atmosphere and all about the rotation of the earth, you may still miss the radiance of the sunset. There is no substitute for the direct perception of the concrete achievement of a thing in its actuality. We want a concrete fact with a high light thrown on what is relevant to its preciousness.

"What I mean is art and aesthetic education . . . What we want is to draw out habits of aesthetic apprehension."[2]

The musician is in a fortunate and unique position to reconcile aesthetic value and scientific measure. He is the only one who can do it, and he can do it precisely. For there is no other field which provides an exact equation between value and measure, like octave $= 1/2$.

A simple experiment on the monochord can prove the point. Ask anyone to divide the given string in half. The first typical reaction will be to ask for a measuring tape, that is, to lean exclusively on a quantitative tool. When the request has been denied, the next reaction is to rely on one's eyes to find the middle. This approach is qualitative, for it calls on the optical sense; but it is not exact and subject to obvious error. Only a musical person, one who takes his senses and values seriously *and* knows what an octave is, will find a quick and infallible solution. He stops the string so that the new pitch is an octave above the old, and the tones sounded by each half are identical; and the *only* spot at which he will succeed will be the *exact* middle.

Is it surprising to a scientific or practical mind that the musician, of all people, has exact tools at his command—much more exact than those of the visual person? Moreover, this precision is directly discernible by our sense apperception without the support of intellectual or mechanical devices.

Our society is accustomed to taking measurements more seriously than values. The following story appeared in the magazine *The New Yorker* some

[2] Alfred North Whitehead, *Science and the Modern World* (New York, 1925), p. 199.

years ago:[3] "Two lady musicians who have lately been trying to conquer Manhattan with their dexterity on violas decided the other day that string-and-bow work was a trifle slack here and headed for a vacation at home in Buffalo, Wyoming. They bought a 1940 DeSoto for a hundred and seventy-five dollars, and plunged westward. As it happened, the thing had a speedometer that wouldn't function, and just before the ladies left, a friend of ours asked them how they were going to figure speeds. 'This DeSoto,' said one of them patronizingly, 'hums in B-flat at fifty. That's all we need to know.' Our friend said they made a nice, confident team as they left."

Clearly, the purpose of the story was to amuse the reader. It is, of course, a funny story but only in a society which respects the contrivance of a speedometer far more than the ears of a music teacher. Whoever believes that musical values are at least as real and reliable as a precision dial on a dashboard might laugh about the implied social commentary of the story more than about the intended oddness.

What both the experiment and the story have illustrated is a truth which no mathematician will deny, namely, that every equation may be reversed. By stating that the octave equals 1/2, we have defined the octave by a ratio. To say that 1/2 equals the octave, that is, to identify a measure by a value, is equally correct. This is what the violinist does who finds the right pitch on his string; this is what the two music teachers did who heard the speed of their car. This is what everyone should do who realizes that in the totality of the world, quantity and quality exist on equal terms—neither one the cause, but each one the symbol, of the other.

There have been times when man operated within this totality. The writings of Lao-tse, Plato, St. Augustine, Dante, and others expound the relation of ethics and number, of music and behavior. The architects of early Gothic cathedrals referred to the "octaveness" of a building in order to indicate a particular spiritual quality beside the geometric ratio of 1:2 betweeen side aisle and nave, between width and height of the nave, and between length and width of the transept. These men belonged to a tradition which, in the Western world, has been called "Pythagorean."

In contrast to Pythagoras who integrated quantity with quality, charging

[3] 3 September 1955, p. 22.

number with meaning, modern man seems to be trying to extract quality from mere figures. The examples of the resulting illusion are endless and can be drawn from trivial as well as fundamental issues. Everybody can find his own. A bigger income is automatically equated with a better life. "Our pain killer works twice as fast," and the implication of a qualitative advantage is not challenged by the possibility that too rapid an action by the medicine might prove harmful. A university proudly publicizes an increase in registration as if more students would automatically increase the level of excellence. The New York Philharmonic Society, in the program booklet of 3-6 January 1963, under the heading "Art at Lincoln Center," used the following terms to describe a hanging sculpture which every subscriber has to pass on the way to his seat: "The completed Orpheus and Apollo weighs 5 tons, which is as if 2½ Cadillacs, or 62.5 players from the New York Philharmonic, were simultaneously suspended from the ceiling." When *Time-Life* in New York sent out printed invitations to an exhibition of color photographs of the Sistine Chapel frescoes, one of the first and prominent sentences read: "During four and a half years [Michelangelo] covered 10,000 square feet with 343 colossal figures."

"Bigger and better" is nonsense, as would be "bigger and worse." There is no simple transfer from the world of measure to the world of value. Only tone permits us to reconcile quantity and quality, for it is both, in a particular and precise manner. Much has been written about the unique role of music, not merely among all cultural phenomena but among the arts themselves. One justification for this elevated position of music has been provided by our first experiment on the sonometer. There is no "octave" in the other arts which would immediately and exactly be perceived by any of our senses. The ear alone among the senses, and music alone among the arts, can reveal spontaneously and precisely the harmony of quantity and quality, the totality of measure and value, in a way which satisfies at once intellect and soul.

DIVISION BY 3

We can now take the next step and divide the string by 3. Again, it is impossible to foretell the musical result without hearing it first. All we may anticipate is that the two new pitches gained by the uneven partition of the string will be one octave from each other: the shorter section measures 1/3 of the total string;

the longer, 2/3; hence the resultant interval equals 1/3 : 2/3, or 1/2, or the octave. But we cannot possibly guess or deduce the relation of either of these pitches to the whole string.

With the help of a measuring tape we stop the string at 1/3 of its total length. The shorter section sounds the interval known as the twelfth above the given string. The remaining 2/3 section sounds the fifth above the open string.

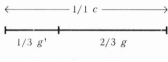

Figure 11

Proceeding as before, we can establish the law that the musical quality of a perfect fifth is bound to the mathematical quantity of 2:3; or, fifth = 2/3. Assuming the identity of the octave for further simplification, we can now forever associate the number 3 with the musical interval of the fifth, just as we have learned to associate the number 2 with the musical interval of the octave.

Here as there, the reciprocity of division and multiplication challenges our curiosity. The same interval that is projected upward by division is projected downward by multiplication:

Figure 12

The result is significant. Out of the force of the number 3 emerge the crucial pitches that stand for the total harmonic space and the maximum tension defined by the cadence: the tonic surrounded by the poles of dominant (1/3) and subdominant (3/1).

Reciprocation should be clearly distinguished from inversion. In music, reciprocation produces an interval in the opposite direction. In the diagram above, for example, 1/3 and 3/1 are reciprocal: the interval is the same, but one of the tones forming the interval has changed (the upward fifth C-G versus the downward fifth C-F). In inversion, on the other hand, the tones remain the same whereas the interval changes (the upward fifth C-G versus the downward fourth C-G).

Because of the special frame set by the octave, outside which the tones repeat themselves, the musician is justified in projecting all intervals into the space of the octave. This operation is called "octave reduction." It is a legitimate operation that reduces intervals to their closest basic distance within the first octave above a fundamental.

In order to project $f_2 = 3/1$ into the first octave above $c = 1/1$—particularly because the experiment of actually multiplying the string by 3 cannot be performed—we "pull it up" two octaves: $3/1 \times 1/2 \times 1/2 = 3/4$. This is the ratio of the perfect fourth, that is, the inverted perfect fifth.

The name "fifth" in this context is as misleading as the name "octave" for the diapason. It so happens that our traditional tone system fills the space created by the relationship of 2:3 with five notes. Other tone systems can easily fill the same space with a different number of tones without changing the basic quality of the outer interval. We need only remember the eight chromatic notes within the fifth in our own tone system to realize the ambiguity of our present nomenclature.

The special quality of the fifth, like that of any other interval, must be heard to be experienced and understood. All one can say about it is that the two tones form a close relationship in which the individuality of each member is more distinct than in the octave. The two tones of the fifth blend but yet preserve their identity.

The particular role of the fifth can be elucidated both phylogenetically and ontogenetically. The first polyphonic singing in history was in parallel fifths, the medieval central-European *organum*. It was preceded only by singing in parallel octaves, which always occurs when men and women, or men and boys, sing the same melody at the same time, and which has never been called "polyphonic." The medieval musicians who had just gained a first insight into the distinctness of the two tones of a fifth happily sang in parallel fifths and rightly considered this performance "polyphonic." The modern musicians who lean toward appreciating the blend of the same two tones just as rightly ban parallel fifths because they are not polyphonic enough.

The strength of the fifth can be heard, ontologically, every time small children sing a melody in chorus. The majority will find the pitch; a few might stray completely; but a good number is likely to sing the melody at a fifth throughout. These children will be surprised to learn that they are "off," that they are not singing the melody as given. To them, the identity of the fifth is stronger than the distinctness. The same singing in parallel fifths can be heard every Sunday morning by a church congregation of average size and ability. In such situations, one hardly ever encounters parallel seconds, for instance, which actually lie closer together and, if distance were the criterion, would be sung more easily and readily than fifths.

An additional factor is supplied by the respective ranges of the human voice. Whereas men and women are one octave apart, high and low voices in either sex are separated by about a fifth (cf. Fig. 32). This factor increases the tendency toward singing in parallel fifths; but only the assumption of norms, such as the ones just developed, explains why the parallel singing does not favor, for example, augmented fourths.

The character of the fifth is described by a reminder of the prevalence of the number 3 in mythology and religion. The symbolism, that is, literally the coincidence of measure and value, spans millennia and continents, as may be here documented by three different writers.

Lao-tse, around 600 B.C., wrote in China: "One has produced Two, Two has produced Three." One of the commentators adds: "These words mean that One has been divided into Yin, the female principle, and Yang, the male principle. These two have joined, and out of their junction came (as a Third) Harmony. The spirit of Harmony, condensing, has produced all beings."

Plato, two centuries later in Greece, wrote: "God made the soul out of the following elements and in this way: Out of the indivisible and unchangeable, and also out of that which is divisible and has to do with material bodies, he compounded a third and intermediate kind of essence . . . He took these three elements . . . and mingled them into one form. . . . When he had mingled them . . . and out of three made one, he again divided this whole into as many portions as was fitting. . ."[4] The division which Plato now proceeds to detail operates exclusively with the first three powers of the first three numbers (1, 2, 3, 4, 8, 9, 27).

A contemporary American university professor, Otto von Simson, writes: "Musical ratios occur in some of the . . . perfect architectural compositions of the thirteenth century. In the southern transept of Lausanne Cathedral (before 1235) the magnificent disposition of the inner wall 'conveys an overwhelming experience of harmony' with the 1:2:3 ratio of its horizontal division.[°] The consonance of the fifth is 'sounded' in the façades of Paris, Strassburg, and York. We . . . find in Chartres itself the realization of the Augustinian aesthetics of measure and number."[5]

DIVISION BY 4

The number 4 is the product of 2 times 2. The musical value of 4 can accordingly be deduced as the octave of the octave. $1/4 = 1/2 \times 1/2 =$ octave above octave. Remember that the mathematical process of multiplying factors always corresponds (logarithmically, as it were) to the musical process of adding intervals. The experiment on the monochord bears out the deduction. The string stopped at $1/4$ of its total length sounds two octaves above the fundamental pitch. By reciprocity, a string 4 times a given length sounds two octaves below it.

A new and important generalization is now possible. The pitch corresponding to any composite number (i.e., a number that is the product of two or more integers greater than 1) can be deduced from its component factors. Any prime number (i.e., a number that is divisible by no number except itself and 1) musically brings to life a new tone.

[4] *Timaeus* 35.

[5] *The Gothic Cathedral* (New York, 1956), pp. 199 f.

[°] Beer, *Die Rose der Kathedrale von Lausanne*, p. 11. For another example of the use of such 'musical' proportions, see Webb, "The Sources of the Design of the West Front of Peterborough Cathedral" (*Archaeological Journal*, LVI, suppl. 1952).

The numbers 1, 2, 3 are all prime numbers. Experiment was necessary to ascertain the corresponding pitches. The number 4 is a composite. By the simple operation of factorizing one can reach the corresponding pitch.

Another generalization becomes inevitable. We know from mathematics that there is no end to prime numbers; they continue into infinity. If each prime number creates a new pitch, must we conclude that there is no end to new pitches—that the number of possible pitches is infinite? Of course, this is the only correct conclusion. In the physical world, the musical vocabulary of different pitches is infinite. In the artistic world, where limitation is a condition *sine qua non*, this infinity will have to be replaced by something manageable. How this has been done, the chapter on musical temperament will expound (cf. pp. 212 ff.). At the moment, however, plain awareness of the existence of the problem suffices.

Knowing the primary interval gained by the division of the string by 4, we may wonder about the pitch of the longer section of the string that has been left over. The length of that section is obviously 3/4. Can we deduce the pitch? In the fraction 3/4, both integers are known to us by experiment. The number 3 is associated with the fifth; and the number 4, with the octave of the octave. Division of the string projects the gained interval upward. Multiplication, being the reverse mathematical operation, projects the interval downward. The fraction 3/4 can thus be translated to mean: up two octaves (1/4) and down a twelfth (3/1), or, a perfect fourth above the given pitch. The string plucked at 3/4 of its length confirms our deduction. We could also have said that the 4th upper partial taken as the new unit now produces its own 3rd lower partial:[6]

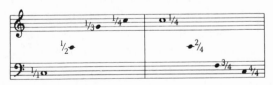

Figure 13

---

[6] "Lower partial" is obviously not a fortunate term. "Partial" is self-evident when relating to "part" of the string but confusing when also applied to a multiple. A new term embodying the idea of multiplication would be a happy solution. Until it is invented, "lower partial" will have to do.

This operation should be remembered in the most general terms. In a fraction expressing the division of a string, the denominator indicates the upward projection of the corresponding musical interval; the numerator, the downward projection.

DIVISION BY 5

Five is a prime number. Dividing the string on the monochord by 5, we must expect a "prime" musical experience, a new pitch. Assisted by a measuring tape, as before, we stop the string at 1/5 of its length and pluck or strike the short section. The pitch we hear is a major third two octaves higher. The musical quality of the major third becomes thus identified with the number 5.

Again, the longer section invites our curiosity, although we are careful to interpret it only as complementary to the basic issue. By now, the answer is quickly reached. The longer section measures 4/5, that is, up a major third and two octaves (1/5), and down two octaves (4/1). The result can be expressed by the equation: 4/5 = major third.

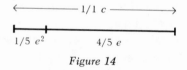

Figure 14

The division of the string into 5 equal parts defines two more, secondary points corresponding to two new pitches. What do we hear when stopping the string at 2/5 of its length? At 3/5 of its length? The pitches can be deduced by the method suggested in the previous section: the denominator projects upward; the numerator, downward. We identify 1/5 as the major third two octaves up and reason that the tone values of 2/5 and 3/5 must lie, respectively, one octave and one twelfth below that tone. Accordingly, 2/5 sounds the major tenth (also intelligible as the octave above the major third, $1/2 \times 4/5$); and 3/5, the major sixth. The correctness of both calculations is verified by what we hear on the monochord.

Figure 15

The reciprocal operation of multiplication yields the following results:

Figure 16

The firm association of the number 5 with the major third prompts us once again to remain aware of the semantic confusion existing between the ratios of musical intervals and the steps of our musical scale. It so happens that the musical fifth is identified with the number 3; and the musical third, with the number 5.

What special quality does the major third, or the number 5, carry into our experience? The least one can say at this moment is that it gives a particular character to the tonal space set off by the preceding numbers: 1, 2, 3, and 4. We recognize this character as belonging to the major mode in the space gained upward by division of the string, and to the minor mode in the space gained downward by multiplication of the string.[7] In the polarity of the major and minor modes, which pervades our music literature, the interval of the third determines the sex. This quality of the number 5 is well reflected by mythology. The pentagram is the symbol of Ishtar, the Babylonian Venus. The Venus temple in Baalbek (to name one of several examples) is pentagonal. Identified with spiritual Love in the Christian religion, the pentagram becomes an amulet, which is occasionally represented in the rose windows of Gothic cathedrals. One may further remember the generating power of the

---

[7] In English, the concept pair "major and minor" has two different musical meanings. One applies to larger and smaller intervals; the other, to the two tonal modes.

quinta essentia ('quintessence') of the alchemists. Johannes Kepler, in *Strena* (which carries the subtitle "Of the Hexagonal Snow"), openly attaches the power of procreation to the occurrence of the number 5 in the world of plants: "If one inquires . . . why most fruit trees and berry bushes develop a blossom precisely according to a pentamerous system, . . . then I say that these things are accounted for by a contemplation of the beauty and particularity of this number, which characterizes the soul of these plants. . . . The fruit from a pentamerous blossom becomes fleshy, as in apples and pears, or pulpy, as in roses and cucumbers, the seed concealed inside the flesh or pulp. But nothing is born from a hexamerous blossom except seed in a dry cavity."[8]

All these symbols and concepts associated with 5 must correspond to something in our soul, which we hear, that is, directly experience, as the major third, the sex of the triad. The interval of the musical third is not merely different from that of the musical fifth: it occupies a more advanced position in the hierarchy of numbers in which 5 comes after 3. Just as the division of the string by 5 is a more complicated operation than the division by 3, so singing in thirds is a more advanced and complicated musical procedure than singing in fifths. This fact is illustrated by the attitude of medieval music theorists who resisted the use of the major third on crucial structural points long after they had admitted the fifth and fourth. The natural hierarchy which places the musical third after the fifth rightly prompted the medieval theorists to make guarded distinctions. To this day, our terminology has preserved the underlying truth by calling the third an "imperfect" consonance as compared to the "perfect" fifth and fourth.

Because the division of the string produces tones in a definite order, intervals differ not merely according to size but just as basically according to value. Singing in parallel thirds was practiced in Northern Europe throughout the late Middle Ages. This practice, called "gymel," corresponds to singing in parallel fifths in France and Italy. The historian who uncovers facts will note both phenomena as manifestations of early polyphony. The musician who commits himself to values will conclude that gymel is not just different from organum but that it also implies a higher degree of tonal awareness.

---

[8] *Gesammelte Werke*, ed. Max Caspar and Franz Hammer (München, 1941), IV, 259-280. Transl. by the authors.

DIVISION BY 6

Six is not a prime number. Hence we can deduce the tones resulting from the division of the string by 6. The musical reading of the equation $1/6 = 1/2 \times 1/3$ tells us that at $1/6$ of the string length we shall hear the octave $(1/2)$ of the twelfth $(1/3)$. So we do after performing the now familiar experiment. Of the secondary nodes produced by the division of the string into 6 parts, all but one repeat earlier results: $2/6 = 1/3$, $3/6 = 1/2$, and $4/6 = 2/3$. A new pitch, however, is created by the longer section of the string divided by 6. We interpret: $5/6 = 1/6 \times 5/1$; or up a fifth plus two octaves, and then down a major third plus two octaves; or the major third below the perfect fifth; or the minor third.

Figure 17

The tone values created by multiplication by 6 can be deduced as before. They can even be produced on the monochord if we take as a starting point, by using octave identity, not the whole string, but a fraction of the string (c^3 at 15 cm, for instance). The following diagram shows that the reciprocal operations of division and multiplication correspond to reciprocal musical experiences: the major triad on one side, and the minor triad on the other.

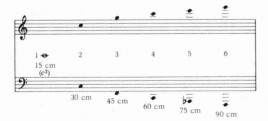

Figure 18

THE SENARIUS

The first six numbers are known as the senarius. There is a special formative power inherent in the senarius—a force that sets limits and thereby shapes the given elements. The pulling-apart initiated by the different forces of 2 and 3 is reconciled at 6. The quality of the senarius—not the quantity now—has left an imprint on the cosmos as much as on our thinking. Crystallography operates primarily with ratios based on the senarius. Snowflakes that deviate from the norm of the hexagon are rare exceptions. The senarius becomes manifest again and again in affinity calculations of chemical elements, in chromosome numbers, in plant structure, et cetera. The numbers of faces, edges, and vertices of the five regular polyhedrons, which are the perfect forms in three-dimensional space, are all determined by senaric values. But man himself has willingly accepted the strength of the senarius. We need only remember the division of the circle in 360 degrees with all its astronomic concomitants: 60 minutes, 24 hours, 30 days of the month, 12 months, 12 signs of the zodiac. These are all ratios based on the senarius; for we understand the senaric values to extend not only over octave identities (1-2-4-8-16..., 3-6-12-24-48..., and 5-10-20-40-80...) but also over products of senaric values (e.g., $9=3\times3$, $15=3\times5$, $24=2\times3\times4$, etc.). Let us also recall our duodecimal system of relating inches to a foot—an old natural measure when compared to the mechanical device of the centimeter. Antique architecture abounds in examples: the 6 columns determining the shape of the Poseidon temple in Paestum; the 6 steps of the oldest Egyptian pyramid at Sakkara, which was built three thousand years before Christ. The element of 6 pervades mythology and all recorded religions. The 12 tribes of the Hebrews or the 12 disciples of Christ represent values as much as numbers. The very formation of the world was determined by the 6 days in which God completed his work.

In music, the senarius (as our experiments on the monochord have shown) is responsible for the creation of the major and minor chords, that is, for the triad as a unified formation and for the polarity of the major and minor modes. This accomplishment suffices to convey an initial idea of the shaping value of the number 6. One should take note, however, that the numbers of the senarius define the basic norms of other musical elements as well. Rhythm and form, for instance, can hardly be understood without reference to the underlying principles of binary and ternary organization.

Exercises

1. Let $1 = c$. What pitches correspond to the following fractions and multiples of the string length: 5/9, 9/16, 3/25, 4/25, 5/27, 8/27, 8/45; 15/4, 25/4, 24/5, 45/8, 81/16?

2. Let $1 = c$. At what fraction of the string lie the tones f-sharp, g-sharp, a-flat, b-flat, b, d^1, a^1, e^2? (Proceed, as directly as possible, by fifths and major thirds, up and down.)

3. Let a string of 120 cm correspond to c. What tones lie at at 10 cm, 12 cm, 15 cm, 24 cm, 25 cm, 27 cm, 32 cm, 45 cm, 80 cm, 81 cm, 90 cm, 96 cm?

4. Let a string of 120 cm correspond to c. Where must it be stopped to sound the tones e-flat, g-flat, a, a-sharp, d^1-sharp, a^1-flat, b^1-flat, d^2?

5. Compute the fractions for all tones of the major scale.

6. Compute the fractions for all tones of the minor scale.

7. Compute the fractions for all tones of the ascending chromatic scale, using exclusively sharps.

8. Compute the fractions for all tones of the descending chromatic scale, using exclusively flats.

9. At what fraction of the string ($c = 1$) lies g-sharp as reached (with eventual octave reduction) by a succession of ascending (a) fifths, (b) major thirds, and (c) fifths and major thirds?

DIVISION BY 7

The special force of the senarius was responsible for temporarily interrupting the flow of our basic experiment of dividing the string. The halt was justified by the consolidating accomplishment of the senarius. The emergence of the triad in music necessitated various reflections. But curiosity, if nothing else, prompts us to divide the string by integers above 6. How high can we go beyond the natural limit of the senarius? What will be the musical results?

Stopping the string at 1/7, we hear a tone that "does not fit" the three tones of the triad gained before by the senarius. It is a minor seventh, but a very flat one. The exact location of this tone on the string can be found easily if we bring this seventh down two octaves to the span of the first octave ($1/7 \times 4/1 = 4/7$). The remainder of the string ($3/7$) will have to sound the perfect fourth above it ($3/7 : 4/7 = 3/4$)—an e^1-flat which, in relation to the senaric system, is as flat as the b-flat at $4/7$.

There is something disquieting about the discovery of a tone that is practically not usable. It stands outside the common language. The tension created by the transgression of the senarius is strongly experienced by the musician listening to the sound of the small seventh, and by man in general coping with the value of the number 7.

For the musician, the seventh affects the perfect equilibrium of the triad by confirming its dynamic potential. The disconcerting character of the seventh admits of no repose. It carries onward. The seventh calls for a transition into another sphere—a statement that the professional musician, in his technical jargon, expresses by saying that this flat seventh demands some kind of resolution. Here lies the reason for the manner in which the seventh has been treated in the theory of harmony and in the practice of the masters.

A sense of ambivalence has generally been attributed to the number 7; and both the discomfort of "not-fitting" and the promise of renewal are amply reflected by folklore and mythology. There are 7 days of mourning, 7 deadly sins, 7 angels of the apocalypse carrying 7 plagues, the animal with 7 horns and eyes, and many other disturbing manifestations. The Revelation of John abounds to such an extent in signs based on 7 that the Bible critic Franz Boll has spoken of a "tyranny of the number 7." But there are also 7 cardinal virtues, 7 seas, 7 wonders of the world, a 7th heaven, 7 guardian angels, and 7 dwarves in the fairy story. The turn to something new, the resolution, is manifested by the sabbath (now the Sunday of our week)—the 7th day of the week which differs in quality from the creative action of the preceding 6. The 8th day renews the cycle, just as the 8th tone of the diatonic scale returns to the octave. If we learn that the Biblical Hebrew word for exorcising has the same root as the word "seven," we can hardly deny the actuality of "seven" as a spiritual force and special form.

Exercises

1. What pitches lie at 2/7, 3/7, 4/7, 5/7, and 6/7 of the total string length?
2. Do all these pitches have a senaric relationship to one another?
3. Name some other pitches to which 3/7, for example, has a senaric relationship.
4. Name some ratios and the corresponding tones to which 3/7, for example, does not have a senaric relationship.

DIVISION BY NUMBERS ABOVE 7

Numbers run into infinity—a fate from which we must preserve our practical experiment. By now we have gathered enough material to be able to pursue the possibilities theoretically. Whenever a number can be factorized, the musical result can be equated with the mathematical product. The logarithmic nature of intervals must be kept in mind: the addition of intervals corresponds to the multiplication of factors. The particular octave range can always be read off the powers of 2 between which a tone happens to lie. Because we call that octave the "first" that lies between 2^0 and 2^1, the octaves take their signature from the power of 2 that defines the upper limit. For instance, 1/9 lies between $1/2^3$ and $1/2^4$, hence within the fourth octave. To pick another number at random, 1/83 lies between $1/2^6$, and $1/2^7$, hence within the seventh octave.

Let us develop the musical results of some further divisions.

$1/8 = 1/2 \times 1/2 \times 1/2 = $ octave $+$ octave $+$ octave.

$1/9 = 1/3 \times 1/3 = $ twelfth $+$ twelfth.

$1/10 = 1/5 \times 1/2 = $ major third, up three octaves.

$1/12 = 1/3 \times 1/2 \times 1/2 = $ perfect fifth, up three octaves.

1/11 involves a prime number. An experiment on the monochord is necessary. The corresponding pitch, in relation to a fundamental *c* sounds a very flat, "out-of-tune," f^3-sharp three octaves up. We have realized earlier (p. 24) that each prime number will produce a new pitch. Now this insight is increased by the additional realization that no prime number from 7 upward can possibly "fit," that is, be musically usable in, a system based on the reign of the senarius. The limiting and hence formative power of the senarius receives in this negative manner an additional confirmation. The precise tone value of each prime number must be heard on the monochord. There is a way, however, of ascertaining the approximate pitch by inference from the data gained thus far. Just as every prime number above the senarius (actually above 5) lies between two composites, so the corresponding tone lies between two ascertainable pitches. Investigating in this manner the division of the string by 11, we could say that the pitch has to lie between 1/10 and 1/12, or between the major third and the perfect fifth, three octaves up. Our inference can even be more precise. If we take an overall view of the tones gained by consecutive divisions of the string, we must notice the shrinking distance between each successive pair of tones.

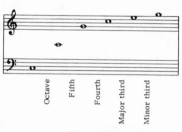

Octave Fifth Fourth Major third Minor third

Figure 19

This observation may be reworded: each tone is slightly closer to its upper than to its lower neighbor. Accordingly, 1/11 will lie closer to the perfect fifth than to the major third. In relation to c, 1/11 is between e^3 and g^3, closer to g^3 —a conclusion borne out by the more precise experiment on the monochord.

The tones that are not usable in our tone system have no name. We can describe them only by reference to the nearest known tone. The signs $\vee$ and $\wedge$ have been used to indicate that a pitch is, respectively, lower or higher than the reference tone. The tone at 1/11 of a string tuned c may accordingly be identified as f^3-sharp$^\vee$.

Similarly, we can work out a few more divisions of the string. The limiting number, which is, of course, arbitrary, is called the "index." Below are all pitches up to the index 16, that is, those obtained when we divide the string successively by the numbers 2 to 16. The pitch c is a convenient assumption for the total string length:

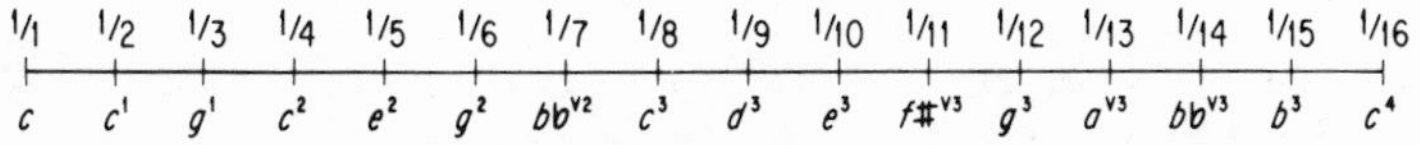

Figure 20

We can enlarge this diagram by registering, in the opposite direction, the multiplications of the string:

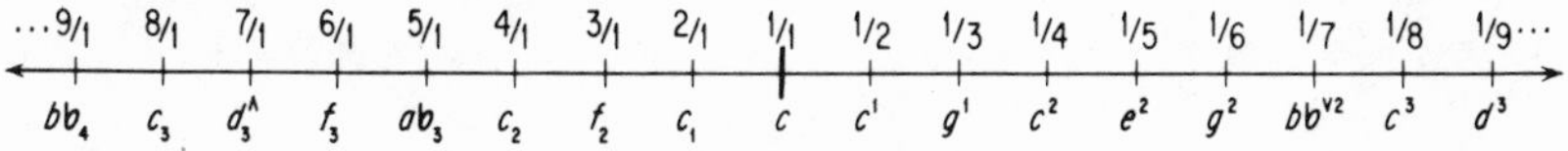

Figure 21

Exercises

1. Divide the string ($c = 1$) into 12 equal parts. What are the pitches that lie at $1/12$, $2/12$, $3/12$, etc.?

2. Let $1 = c$. What pitches correspond to the following fractions and multiples of the string length: $3/11$, $7/11$, $10/11$, $4/13$, $6/17$; $11/7$, $13/9$, $19/5$?

THE PYTHAGOREAN TABLE

Our detailed investigation of the division and multiplication of the string by the integers of the senarius has paid some attention to the secondary results, that is, to the tones reached by multiples of the primary fraction. Thus in the case of the division by 5, for example, we recognized the existence, not only of $1/5\ e^2$, but also of $2/5\ e^1$, $3/5\ a$, and $4/5\ e$. The following graphic representation, which transforms the linear arrangement (cf. Fig. 21) into a system of coordinate axes, is useful and revealing:

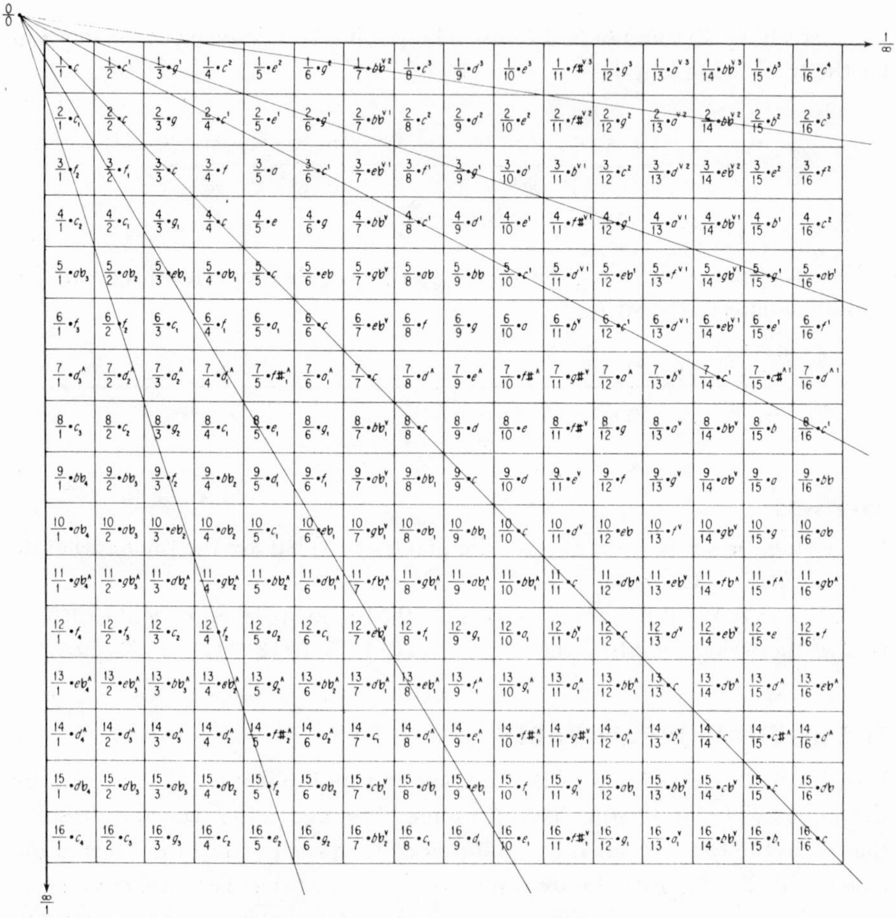

$\frac{0}{0}$ $\frac{1}{\infty}$

$\frac{1}{1}$·c	$\frac{1}{2}$·c¹	$\frac{1}{3}$·g¹	$\frac{1}{4}$·c²	$\frac{1}{5}$·e²	$\frac{1}{6}$·g²	$\frac{1}{7}$·bb^{v2}	$\frac{1}{8}$·c³	$\frac{1}{9}$·d³	$\frac{1}{10}$·e³	$\frac{1}{11}$·f♯^{v3}	$\frac{1}{12}$·g³	$\frac{1}{13}$·a^{v3}	$\frac{1}{14}$·bb³	$\frac{1}{15}$·b³	$\frac{1}{16}$·c⁴
$\frac{2}{1}$·c₁	$\frac{2}{2}$·c	$\frac{2}{3}$·g	$\frac{2}{4}$·c¹	$\frac{2}{5}$·e¹	$\frac{2}{6}$·g¹	$\frac{2}{7}$·bb^{v1}	$\frac{2}{8}$·c²	$\frac{2}{9}$·d²	$\frac{2}{10}$·e²	$\frac{2}{11}$·f♯^{v2}	$\frac{2}{12}$·g²	$\frac{2}{13}$·a^{v2}	$\frac{2}{14}$·bb²	$\frac{2}{15}$·b²	$\frac{2}{16}$·c³
$\frac{3}{1}$·f₂	$\frac{3}{2}$·f₁	$\frac{3}{3}$·c	$\frac{3}{4}$·f	$\frac{3}{5}$·a	$\frac{3}{6}$·c¹	$\frac{3}{7}$·eb¹	$\frac{3}{8}$·f¹	$\frac{3}{9}$·g¹	$\frac{3}{10}$·a¹	$\frac{3}{11}$·b¹	$\frac{3}{12}$·c²	$\frac{3}{13}$·d^{v2}	$\frac{3}{14}$·eb²	$\frac{3}{15}$·e²	$\frac{3}{16}$·f²
$\frac{4}{1}$·c₂	$\frac{4}{2}$·c₁	$\frac{4}{3}$·g₁	$\frac{4}{4}$·c	$\frac{4}{5}$·e	$\frac{4}{6}$·g	$\frac{4}{7}$·bb^v	$\frac{4}{8}$·c¹	$\frac{4}{9}$·d¹	$\frac{4}{10}$·e¹	$\frac{4}{11}$·f♯^{v}	$\frac{4}{12}$·g¹	$\frac{4}{13}$·a^{v1}	$\frac{4}{14}$·bb¹	$\frac{4}{15}$·b¹	$\frac{4}{16}$·c²
$\frac{5}{1}$·ab₂	$\frac{5}{2}$·ab₁	$\frac{5}{3}$·eb₁	$\frac{5}{4}$·ab₁	$\frac{5}{5}$·c	$\frac{5}{6}$·eb	$\frac{5}{7}$·gb^v	$\frac{5}{8}$·ab	$\frac{5}{9}$·bb	$\frac{5}{10}$·c¹	$\frac{5}{11}$·d^{v1}	$\frac{5}{12}$·eb¹	$\frac{5}{13}$·f^{v1}	$\frac{5}{14}$·gb¹	$\frac{5}{15}$·g¹	$\frac{5}{16}$·ab¹
$\frac{6}{1}$·f₃	$\frac{6}{2}$·f₂	$\frac{6}{3}$·c₁	$\frac{6}{4}$·f₁	$\frac{6}{5}$·a₁	$\frac{6}{6}$·c	$\frac{6}{7}$·eb¹	$\frac{6}{8}$·f	$\frac{6}{9}$·g	$\frac{6}{10}$·a	$\frac{6}{11}$·b^{v}	$\frac{6}{12}$·c¹	$\frac{6}{13}$·d^{v1}	$\frac{6}{14}$·eb¹	$\frac{6}{15}$·e¹	$\frac{6}{16}$·f¹
$\frac{7}{1}$·d₃^A	$\frac{7}{2}$·d₂^A	$\frac{7}{3}$·a₂^A	$\frac{7}{4}$·d₁^A	$\frac{7}{5}$·f♯₁^A	$\frac{7}{6}$·a₁^A	$\frac{7}{7}$·c	$\frac{7}{8}$·d^A	$\frac{7}{9}$·e^A	$\frac{7}{10}$·f♯^A	$\frac{7}{11}$·g♯^A	$\frac{7}{12}$·a^A	$\frac{7}{13}$·b^A	$\frac{7}{14}$·c¹	$\frac{7}{15}$·c♯^{A1}	$\frac{7}{16}$·d^{A1}
$\frac{8}{1}$·c₃	$\frac{8}{2}$·c₂	$\frac{8}{3}$·g₂	$\frac{8}{4}$·c₁	$\frac{8}{5}$·e₁	$\frac{8}{6}$·g₁	$\frac{8}{7}$·bb^v	$\frac{8}{8}$·c	$\frac{8}{9}$·d	$\frac{8}{10}$·e	$\frac{8}{11}$·f♯^v	$\frac{8}{12}$·g	$\frac{8}{13}$·a^v	$\frac{8}{14}$·bb	$\frac{8}{15}$·b	$\frac{8}{16}$·c¹
$\frac{9}{1}$·bb₃	$\frac{9}{2}$·bb₂	$\frac{9}{3}$·f₂	$\frac{9}{4}$·bb₁	$\frac{9}{5}$·d₁	$\frac{9}{6}$·f₁	$\frac{9}{7}$·ab₁	$\frac{9}{8}$·bb	$\frac{9}{9}$·c	$\frac{9}{10}$·d	$\frac{9}{11}$·e^v	$\frac{9}{12}$·f	$\frac{9}{13}$·g^v	$\frac{9}{14}$·ab	$\frac{9}{15}$·a	$\frac{9}{16}$·bb
$\frac{10}{1}$·ab₃	$\frac{10}{2}$·ab₂	$\frac{10}{3}$·eb₂	$\frac{10}{4}$·ab₁	$\frac{10}{5}$·c₁	$\frac{10}{6}$·eb₁	$\frac{10}{7}$·gb₁	$\frac{10}{8}$·ab₁	$\frac{10}{9}$·bb₁	$\frac{10}{10}$·c	$\frac{10}{11}$·d^v	$\frac{10}{12}$·eb	$\frac{10}{13}$·f^v	$\frac{10}{14}$·gb	$\frac{10}{15}$·g	$\frac{10}{16}$·ab
$\frac{11}{1}$·gb₄^A	$\frac{11}{2}$·gb₃^A	$\frac{11}{3}$·db₂^A	$\frac{11}{4}$·gb₂^A	$\frac{11}{5}$·bb₁^A	$\frac{11}{6}$·db₁^A	$\frac{11}{7}$·fb₁	$\frac{11}{8}$·gb₁^A	$\frac{11}{9}$·ab₁^A	$\frac{11}{10}$·bb₁^A	$\frac{11}{11}$·c	$\frac{11}{12}$·db^v	$\frac{11}{13}$·eb^v	$\frac{11}{14}$·fb^v	$\frac{11}{15}$·f^A	$\frac{11}{16}$·gb^A
$\frac{12}{1}$·f₄	$\frac{12}{2}$·f₃	$\frac{12}{3}$·c₂	$\frac{12}{4}$·f₂	$\frac{12}{5}$·a₂	$\frac{12}{6}$·c₁	$\frac{12}{7}$·eb₁^v	$\frac{12}{8}$·f₁	$\frac{12}{9}$·g₁	$\frac{12}{10}$·a₁	$\frac{12}{11}$·b^v	$\frac{12}{12}$·c	$\frac{12}{13}$·d^v	$\frac{12}{14}$·eb	$\frac{12}{15}$·e	$\frac{12}{16}$·f
$\frac{13}{1}$·eb₄^A	$\frac{13}{2}$·eb₃^A	$\frac{13}{3}$·bb₂^A	$\frac{13}{4}$·eb₂^A	$\frac{13}{5}$·g₂^A	$\frac{13}{6}$·bb₁^A	$\frac{13}{7}$·db₁^A	$\frac{13}{8}$·eb₁^A	$\frac{13}{9}$·f₁^A	$\frac{13}{10}$·g₁^A	$\frac{13}{11}$·a₁^A	$\frac{13}{12}$·bb^v	$\frac{13}{13}$·c	$\frac{13}{14}$·db^v	$\frac{13}{15}$·d^A	$\frac{13}{16}$·eb^A
$\frac{14}{1}$·d₄^A	$\frac{14}{2}$·d₃^A	$\frac{14}{3}$·a₃^A	$\frac{14}{4}$·d₂^A	$\frac{14}{5}$·f♯₂^A	$\frac{14}{6}$·a₂^A	$\frac{14}{7}$·c₁	$\frac{14}{8}$·d₁^A	$\frac{14}{9}$·e₁^A	$\frac{14}{10}$·f♯₁^A	$\frac{14}{11}$·g♯₁^A	$\frac{14}{12}$·a₁^A	$\frac{14}{13}$·b₁^A	$\frac{14}{14}$·c	$\frac{14}{15}$·c♯^A	$\frac{14}{16}$·d^A
$\frac{15}{1}$·db₄^A	$\frac{15}{2}$·db₃^A	$\frac{15}{3}$·ab₃	$\frac{15}{4}$·db₂^A	$\frac{15}{5}$·f₂	$\frac{15}{6}$·ab₂	$\frac{15}{7}$·cb₁	$\frac{15}{8}$·db₁^A	$\frac{15}{9}$·eb₁	$\frac{15}{10}$·f₁	$\frac{15}{11}$·g^v	$\frac{15}{12}$·ab	$\frac{15}{13}$·bb^v	$\frac{15}{14}$·cb¹	$\frac{15}{15}$·c	$\frac{15}{16}$·db
$\frac{16}{1}$·c₄	$\frac{16}{2}$·c₃	$\frac{16}{3}$·g₃	$\frac{16}{4}$·c₂	$\frac{16}{5}$·e₂	$\frac{16}{6}$·g₂	$\frac{16}{7}$·bb₁^v	$\frac{16}{8}$·c₁	$\frac{16}{9}$·d₁	$\frac{16}{10}$·e₁	$\frac{16}{11}$·f♯₁^v	$\frac{16}{12}$·g₁	$\frac{16}{13}$·a₁^v	$\frac{16}{14}$·bb₁	$\frac{16}{15}$·b₁	$\frac{16}{16}$·c

$\frac{\infty}{1}$

Figure 22

This diagram, in a slightly different form with the angle at the apex, was well known to the Greeks. The neo-Platonist Iamblichus called it the "lambdoma" because the shape Λ resembles the Greek letter lambda. Albert von Thimus, in his fundamental work on *Die harmonikale Symbolik des Alther-*

thums, named it the "Pythagorean table."[9] Thimus was primarily concerned with the historical aspects of the musical number symbolism. Basing his research on the table, he succeeded in giving a definitive explanation of Plato's numerical representation of the world soul in the dialogue *Timaeos*, and in connecting it with a long Pythagorean tradition. Thimus and Kayser may be profitably studied by anyone wishing to develop the mathematical properties of the table in its various projections or to follow its symbolic and philosophic implications. We shall limit ourselves to pointing out only a few relevant qualities of the table.

On the legs of the table, at right angle to each other, we discern two series. One represents the division of the string; the other, the multiplication. Each series begins at 1 at the apex. Each progresses to infinity—one to the infinitely small ($1/\infty$), the other to the infinitely large ($\infty/1$). The space between both legs of the table is filled by interpolations, which attest to the independent life of each number of the series. Every tone is located at the intersection of two polar series. For instance, 3/5 lies at the intersection of the series initiated by 1/5 and 3/1.

Remembering that division of the string eventuates in the major triad, and that multiplication of the string eventuates in the minor triad, we may call the two series "major" and "minor." We can read off the table that major and minor participate in varying proportions in the production of tones. They cancel each other in the diagonal 1/1, 2/2, 3/3 . . . n/n. If lines are drawn connecting identical tones (e.g., 1/1 *c*, 2/2 *c*, 3/3 *c* . . .; 2/3 *g*, 4/6 *g*, 6/9 *g* . . .; 3/4 *f*, 6/8 *f*, 12/16 *f* . . . ; 1/3 g^1, 2/6 g^1, 3/9 g^1), all of these lines will meet beyond the apex 1/1 in a point logically designated as 0/0. These lines, of which there is an unlimited amount growing with the index of the table, are called "identity rays."

The consequences of the Pythagorean table for the theory of harmony are numerous and remarkable. Beyond the immediate significance, the table contains philosophic implications that transcend the purely historic interest. Beside being a veritable key to Platonic-Pythagorean philosophy, as Thimus and Kayser have shown, the concepts suggested by the table are not confined to a singular school of thought but have been operative through the Middle Ages

[9] 2 vols. (Köln, 1868-76).

and beyond. The Omega concept of the modern scientist and philosopher Teilhard de Chardin, for instance, is easily equated with, and precisely elucidated by, the 0/0 point of the lambdoma. Nor did Christian theologians fail to recognize the symbolic representation by the table of the Holy Trinity at the 0/0 and 1/1 points ("three in one," "two in one") and of the process of creation and procreation inside the expanding network.

How the Pythagorean table may be used as a geometric dividing canon, one of the exercises at the end of this chapter (cf. pp. 39 f.) explicates. Geometric dividing canons were widely used in former times, especially by architects. The reason lay not only in the difficulty of readily finding graduated rulers. Specifically, architects were interested first and above all, not in scale drawings, but in proportioned drawings, for which division by geometrical means served as the perfectly appropriate method. The use of a harmonical dividing canon, in substance related to the Pythagorean table, is documented throughout the Middle Ages up to at least the sixteenth century.[10]

A curious analogy exists between the Pythagorean table and the periodic table of elements. In both cases, a linear series of numbers came to be known. The series was felt to be significant, but its hidden law was not discovered before the application of an "artificial" arrangement by means of interpolation resulting in a planimetric development. Both the lambdoma and the periodic table are basically projections of the structure of our mind, regardless of the possibilities of their application to the outer world. As such, they may be called "icons" in the truest and purest sense: images of forms of our spirit. The periodic table of the chemist has no reality in the ordinary sense, for nowhere are the elements found packed together in the order determined by the table. We deal here with an inner form concept which, first, has an intrinsic value; second, proves to be applicable and thereby demonstrates an attunement between the inner and outer worlds; and, third, shows the existence of a higher kind of order that remains invisible under ordinary circumstances. The same remarks hold true of the Pythagorean table, the differences lying primarily in the material contents—musical tones instead of chemical elements. Even so, and notwithstanding the infancy of relevant contemporary studies, the applica-

[10] Hans Kayser, *Ein harmonikaler Teilungs-Kanon* (Zürich, 1946).

bility of the special form concepts presented by the lambdoma to certain phenomena of the outer world has proven helpful and true.

Exercises

1. Draw a Pythagorean table and isolate the square lying between the indexes 4 and 8. Within that square, find the following chords:

    a. The diminished triad lying on the cross 5-7 with the center on 6.

    b. The two seventh chords of which this diminished triad is the common portion.

    c. All D^7 and S_7 chords lying between the indexes 4 and 7.

    d. The two diminished seventh chords:

Figure 23

Connect the tones that form the various chords with pencil lines of various colors. Observe the genesis of each chord, which elucidates the actual structure of each chord and provides clues to the possible resolutions and connections.

2. How to use the Pythagorean table as a dividing canon is shown in Fig. 24. Draw a rather large table (about 2" or 4 cm for the side of each square). Fix the index at will but (for practical reasons) at not less than 9. Draw a number of identity rays. Trim the paper along the 0/0 line, that is, exactly one-half square behind the original series 1/1, 1/2 ... and 1/1, 2/1 .... Slide the table under the strings so that the upper end of the sheet is flush against the permanent metal bridge. The line to be divided is represented by the string length from the metal bridge (the horizontal 0/0 axis) to the intersection with the diagonal (the 1/1 identity ray). The string may lie over any vertical line of the table, because the rays are proportionately spaced at any point; but the nearer the monochord string to the right edge of the table, the more exact the obtained results, because of the greater absolute distance from one ray to the next. In the chosen position, tack the table to the monochord. At the point where the diagonal cuts the string, place one of the movable bridges (in Fig. 24 this point is marked by a circle around it). If you wish, you may tune the

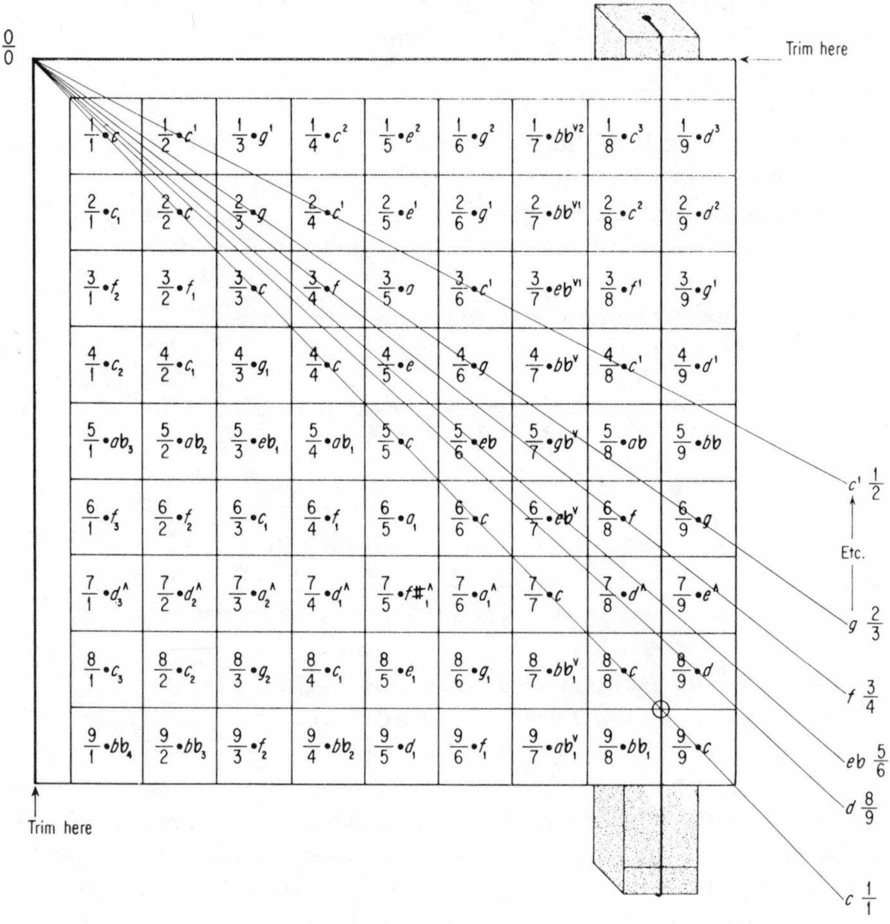

$\frac{0}{0}$

Trim here

$\frac{1}{1}\bullet c$	$\frac{1}{2}\bullet c^1$	$\frac{1}{3}\bullet g^1$	$\frac{1}{4}\bullet c^2$	$\frac{1}{5}\bullet e^2$	$\frac{1}{6}\bullet g^2$	$\frac{1}{7}\bullet bb^{v2}$	$\frac{1}{8}\bullet c^3$	$\frac{1}{9}\bullet d^3$
$\frac{2}{1}\bullet c_1$	$\frac{2}{2}\bullet c$	$\frac{2}{3}\bullet g$	$\frac{2}{4}\bullet c^1$	$\frac{2}{5}\bullet e^1$	$\frac{2}{6}\bullet g^1$	$\frac{2}{7}\bullet bb^{vi}$	$\frac{2}{8}\bullet c^2$	$\frac{2}{9}\bullet d^2$
$\frac{3}{1}\bullet f_2$	$\frac{3}{2}\bullet f_1$	$\frac{3}{3}\bullet c$	$\frac{3}{4}\bullet f$	$\frac{3}{5}\bullet a$	$\frac{3}{6}\bullet c^1$	$\frac{3}{7}\bullet eb^{vi}$	$\frac{3}{8}\bullet f^1$	$\frac{3}{9}\bullet g^1$
$\frac{4}{1}\bullet c_2$	$\frac{4}{2}\bullet c_1$	$\frac{4}{3}\bullet g_1$	$\frac{4}{4}\bullet c$	$\frac{4}{5}\bullet e$	$\frac{4}{6}\bullet g$	$\frac{4}{7}\bullet bb^v$	$\frac{4}{8}\bullet c^1$	$\frac{4}{9}\bullet d^1$
$\frac{5}{1}\bullet ab_3$	$\frac{5}{2}\bullet ab_2$	$\frac{5}{3}\bullet eb_1$	$\frac{5}{4}\bullet ab_1$	$\frac{5}{5}\bullet c$	$\frac{5}{6}\bullet eb$	$\frac{5}{7}\bullet gb^v$	$\frac{5}{8}\bullet ab$	$\frac{5}{9}\bullet bb$
$\frac{6}{1}\bullet f_3$	$\frac{6}{2}\bullet f_2$	$\frac{6}{3}\bullet c_1$	$\frac{6}{4}\bullet f_1$	$\frac{6}{5}\bullet a_1$	$\frac{6}{6}\bullet c$	$\frac{6}{7}\bullet eb^v$	$\frac{6}{8}\bullet f$	$\frac{6}{9}\bullet a$
$\frac{7}{1}\bullet d_3^\wedge$	$\frac{7}{2}\bullet d_2^\wedge$	$\frac{7}{3}\bullet g_2^\wedge$	$\frac{7}{4}\bullet d_1^\wedge$	$\frac{7}{5}\bullet f\#_1^\wedge$	$\frac{7}{6}\bullet a_1^\wedge$	$\frac{7}{7}\bullet c$	$\frac{7}{8}\bullet d^\wedge$	$\frac{7}{9}\bullet e^\wedge$
$\frac{8}{1}\bullet c_3$	$\frac{8}{2}\bullet c_2$	$\frac{8}{3}\bullet g_2$	$\frac{8}{4}\bullet c_1$	$\frac{8}{5}\bullet e_1$	$\frac{8}{6}\bullet g_1$	$\frac{8}{7}\bullet bb^v_1$	$\frac{8}{8}\bullet c$	$\frac{8}{9}\bullet d$
$\frac{9}{1}\bullet bb_4$	$\frac{9}{2}\bullet bb_3$	$\frac{9}{3}\bullet f_2$	$\frac{9}{4}\bullet bb_2$	$\frac{9}{5}\bullet d_1$	$\frac{9}{6}\bullet f_1$	$\frac{9}{7}\bullet ab^v_1$	$\frac{9}{8}\bullet bb_1$	$\frac{9}{9}\bullet c$

$c^1\ \frac{1}{2}$

Etc.

$g\ \frac{2}{3}$

$f\ \frac{3}{4}$

$eb\ \frac{5}{6}$

$d\ \frac{8}{9}$

$c\ \frac{1}{1}$

Trim here

Figure 24

string so that the portion stretching between the 0/0 axis and the bridge actually sounds c; but this procedure is unnecessary because the division is based
on relative rather than absolute pitch. Each intersection of the string with
an identity ray now corresponds exactly to the division of the string indicated
by that ray. If you stop the string at these intersections by moving the bridge
along it, you can hear the tones indicated by the various rays.

3 | Wave Properties

SOUND WAVES

The vibration of the monochord string has a certain frequency, which can be measured in time. The vibration also relates to the length of the string, which can be measured in space. This spatial property of a vibration is generally designated in physics by the term "wavelength." It is the distance from any one point on a wave to the next point at which, at the same instant, there is the same phase. For example, the wavelength is the distance from maximum compression to maximum compression in an air wave; or from the crest of one ocean wave to the crest of the next. To the musician, visualizing the monochord, wavelength may be correlated with string length. We readily see that wavelength and frequency are reciprocal: the longer the wave, the lower the frequency; and the shorter the wave, the higher the frequency. To transform a ratio between wavelengths (i.e., string lengths) into one between frequencies, we need merely turn the ratio around: 2/3 string = 3/2 frequency. The product of wavelength and frequency is constant.

Every wave implies motion. The vibratory motion in a sound wave may be either transverse or longitudinal.

Transverse motion is well demonstrated by the movement of a long, heavy cord, fastened at one end to, say, a door knob and held at the free end by your hand so that the cord will be loosely stretched. A rapid up-and-down motion of your hand, just once, will send a wave through the cord. The wave travels toward the door in a manner resembling the motion of a water wave. Whereas the wave actually travels, the cord does not. The particles of the cord perform an up-and-down motion: this oscillation at right angles to the direction of the wave gives the transverse wave its name. The wave is reflected from the fastened end and travels back toward you. The reflection is a negative of the outgoing motion: crests become troughs, and troughs become crests.

Because of its nature, a wave of this kind is also called a "traveling wave." If you now agitate the cord continuously, it will begin to oscillate as a whole, like a vibrating string, at a certain speed of your hand. By increasing the speed of the hand motion, you can cause the cord to oscillate in higher modes—1/2, 1/3, etc., corresponding to octave, fifth, etc. This new way of moving can be thought of as a combination of the outgoing and reflected waves. Notice that the wave does not travel any more: it has become a "standing wave."

A longitudinal wave is less conveniently demonstrated. The behavior of beads strung horizontally conveys a fair idea when you hit the end bead. The bead will hit its neighbor, which will hit the following bead, and so on to the other end. What you observe is a wave of alternating compressions and rarefactions traveling through the row. (You can make similar observations playing with one of the loose coil springs which one frequently finds in toy stores.) Like the transverse wave, the longitudinal wave itself travels while the particles (the beads) oscillate in the path of the wave. But here the oscillation is to-and-fro instead of up-and-down, in the direction of the wave instead of at right angles to it.

Longitudinal standing waves form analogously to transverse standing waves, but they are difficult to demonstrate. One may observe them indirectly by their action on a light powder or fine sand in the "Kundt tube," where the powder or sand will gather in little heaps at the points corresponding to the nodes of a standing wave that has been set in motion inside the tube.

The velocity of a sound wave in air, which is the normal transmitter from the source to the ear, has been experimentally determined as about 1100 feet at 32° Fahrenheit (or about 330 meters at 0° centigrade) per second. This velocity is equivalent to about 750 miles per hour. In warmer air, sound travels a bit faster (about 1140 feet at 80° Fahrenheit), because heated air is less dense and therefore exhibits less inertia than cold air. The velocity is also increased by humidity, because a molecule of water is less massive, and hence has less inertia, than a typical molecule of air.

A precise relationship can be set up connecting velocity, frequency, and wavelength. The product of the last two, we remember, is constant (cf.p. 41).

We now know that each wave travels at a velocity of about 1100 feet per second. At a frequency of 1 vibration per second, the wavelength would take up this entire distance. At a frequency of 2 vibrations per second, there being two waves in the same distance, the wavelength would be one-half of it. At a frequency of 3, three waves have to be accommodated within 1100 feet, and the wavelength is accordingly one-third of 1100. In general terms, the situation is expressed by the equation: wavelength = velocity/frequency; or: wavelength × frequency = velocity.

The velocity of sound is well within the reach of our concrete imagination and very slow when compared to that of light. This discrepancy may be annoying in large auditoriums, where the visual impression of the performer on the stage reaches the spectator noticeably ahead of the acoustical impression. Each cymbal crash in the march from Tchaikovsky's *Pathétique* symphony, for instance, is visible to a listener in the last row of the Hollywood Bowl one full beat before it becomes audible. If this listener knows that the tempo of the movement is circa ♩ = 144, he can readily compute the distance between him and the orchestra across the Hollywood Bowl to be about 475 feet (the time lapse multiplied by the velocity, or 60/144 × 1140).

For all practical purposes, the velocity is the same for all tones, regardless of whether they are high or low, loud or soft, sung or produced by different instruments. This fact is fortunate, to say the least, for the rendition of music; for the various tones of a composition reach the ear in the same order in which they are produced, whether high or low, loud or soft, bowed or blown.

Yet, why do we hear the bass notes of a band marching away from us when the higher notes all seem to have vanished? The explanation has nothing to do with the velocity at which the sound travels toward us but rather with the greater ability of long waves (which carry the low tones) to bend around obstacles, such as houses. Short waves cannot circumvent these same obstacles and are absorbed by them. Mozart made good use of this phenomenon in the cemetery scene of *Don Giovanni*, where in the retreat of the two men from the tomb the higher voice of Don Giovanni disappears two measures before the lower voice of his servant. The sunset supplies a visual counterpart: the long waves on the red side of the spectrum are not so easily absorbed by dust particles as are the shorter waves on the violet side; and as a result, sunsets are generally red and not violet.

Why does the motor pitch of an airplane flying overhead, or of a car driving toward and then past you on a highway, seem to rise as it approaches you at high speed, and then to drop as it moves away? Here, the velocity of sound really provides the explanation. The rapid motion toward the observer crowds the waves, which accordingly reach the observer's ear in greater number per time unit. The inverse happens when the airplane or car is rapidly moving away from the observer. The phenomenon is known as "Doppler effect," in honor of the Austrian physicist Christian Johann Doppler (1803-1853) who first described it.

Exercises

1. The *Te Deum* by Berlioz is to be performed in a large church. Choir and orchestra are placed in the apse; the organ is located at the opposite end of the nave, 290 feet away. At a tempo of $\quad = 60$, what is the time lag (expressed in a note value) of the orchestra for a hearer placed near the organ?
2. During a thunderstorm, compute the approximate distance of the storm from your location by the time lag between the perception of the flash (which can be assumed to be instantaneous) and that of the noise.
3. Some places are remarkable for sharp echoes (forest fronts, mountainsides, walls). In such a location, produce a loud and short sound. Then compute your approximate distance from the reflecting surface.
4. What is the wavelength in air of the tone *a*, which has a frequency of 440?

RESONANCE

The vibrations of a body can reach out and set off vibrations in another body. This fact is called "resonance" (from the Latin verb *resonare*, 'to return a sound'). If the two bodies have identical frequencies, we speak of "free" resonance or "sympathetic vibration." If the two bodies have different frequencies, we speak of "forced" resonance.

The latter case (which in the strictest sense should not be called "resonance") occurs when the primary vibration is transmitted by force, as it were, to some other body. The vibrating string of a monochord or violin, for instance, forces the wood of the soundboard to vibrate along regardless of the respective frequencies; and as a result, the tone is intensified and enriched. Without the reinforcement provided by the resonance of the soundbox, the tone of the vibrating string would be audible but very weak.

Free resonance or sympathetic vibration should be kept strictly distinct from forced resonance. It can be easily demonstrated on the monochord. Tune some of the strings to exactly the same pitch, others a bit off. Then pluck or strike a string and notice that the other strings of the same pitch—and only those—will pick up the vibration. They will continue their own vibration even after the plucked or struck string has been stopped. The intensity of the resonance will vary with certain factors, primarily with the intensity of the original vibration. Little paper riders set on the strings will make the phenomenon visible by falling off the resonant strings of the same frequency and remaining astride the others.

Other experiments on free resonance offer themselves readily. Take two tuning forks of the same pitch. Strike one, and the other will resound. Or hold the pedal of a piano to remove the dampers from the strings, and then sing or play a note near the piano: you will hear the sound returned by the string of the corresponding pitch. A radio receiver might be thought of as a kind of powerful resonator: by setting the dial, we "tune" the instrument, that is, we adjust the frequency of electrical vibrations at various points in the receiver to coincide with the frequency of the sender so that a particular broadcast, meeting an instrument sympathetic to its vibrations, is made to "re-sound." A less controlled situation is presented by objects in a room that quiver audibly in response to a certain pitch on the piano. In a concert hall of recent construction a loose metal plate on the ceiling started to vibrate audibly whenever an *a*-flat was produced by the orchestra on the stage.

The important factor in all these phenomena of free resonance is the readiness of a body to respond to a frequency identical with its own potential frequency. Without this identity, there would be no resonance. The general validity of this basically musical idea has long been recognized by the sciences. The acoustical term "resonance" has been appropriated by physics and chemistry, as a mere glance at the dictionary reveals. The idea of resonance may be more than a mere metaphor when used to describe sympathies or antipathies between people.

Exercises

1. Hold a vibrating tuning fork in your hand until the tone becomes inaudible. Then set the tuning fork immediately on a table top, and the tone will reappear. Is this an example of free or of forced resonance?

2. Having opened the lid of a piano and freed the strings from the dampers by depressing the right pedal, sing a tone into the open instrument. The corresponding string will resound. Determine its pitch by trying to find the key acting on that string.

3. Keep your ears open for the rattling of any object in your room in resonance to a certain tone produced by radio, phonograph, piano, voice, etc. Determine the critical pitch. Change, if possible, the frequency of the vibration of the rattling object by weighting it, for instance, with putty or water. Now observe that it no longer rattles in response to the original pitch.

THE OVERTONE SERIES

We have divided the vibrating string in controlled experiments. Nature performs the same divisions spontaneously. It divides any vibrating body that produces a pure musical tone according to the whole-number series. As a result, the phenomenon which we recognize as one tone is in reality a complex of many partial tones. The partial tones above the fundamental are also called "overtones." Their existence is a manifestation of a natural law, which can be demonstrated audibly by the following few experiments.

While bowing a string on the monochord (or on any stringed instrument), let a finger of the free hand slide lightly up and down the string. You will notice that only certain spots readily yield definite tones, which are known to every string player as "harmonics." Ask a horn player or a trumpeter to blow into his instrument and to produce different pitches merely by his embouchure without using any valves, keys, or slides. He will succeed in producing only certain tones—always the same ones, and with increased technical struggle as these tones pass beyond a certain height. On the piano, remove the dampers from the tones c-e-g by mutely pressing down the corresponding keys, and then strike loudly the low c_2. After releasing and thus stopping the struck c_2, you will hear clearly the sound of the three tones that had not been struck at all. They existed, so to speak, as part of the complex totality of the fundamental tone. Our experiment simply brought them to life by employing the principle of resonance, which we have encountered earlier (cf. pp. 44 ff.).

These three little experiments share not merely the physical background of factual vibrations, but they share identical results. The *only* partial tones produced by the bowing of the string, the blowing of the horn or trumpet, and the resonance of the piano keys are the following (here shown in relation to the fundamental tone c_2:

Figure 25

If the fundamental changes, the partials will, of course, change accordingly. But their relation to the fundamental will always remain constant. In the experiment, the violinist and the trumpeter will be unable to sound any other harmonics in relation to a given fundamental except the ones specified above. The pianist will not hear the partial tones of c_2 by holding down, for instance, the keys $c\sharp$-$e\sharp$-$g\sharp$.

Several observations arise from the study of the overtone series.

First, we notice an exact coincidence of the overtone series with the results of our divisions on the monochord:

Figure 26

This is not an accident but a symptom of the truth that the norms of our psyche correspond with those of the physical world. Man is part of, and not outside, nature. The physicist usually explains the overtone series as a natural fact, and he is, of course, correct from his point of view. The wind blowing through a reed can produce the partial tones without the participation of man. The natural tones heard and observed then lead the physicist to the definition of the numerical law. The physicist can point out that every particle of a vibrating string participates in the fundamental vibration and that a vibrating string will divide itself into any number of identical segments; each such segment produces a corresponding overtone. The musician explains the overtone series as a psychological reality which exists not only in nature but which necessarily agrees with a particular norm of our soul. If this were not so, we would not respond to the particular character of the various tone values. The musician postulates laws of proportion and then hears the corresponding tones as a psychological result. Both the physical and the psychological approaches are justified. But because the overtone series is commonly represented exclusively as a natural phenomenon, the musician must not forget to emphasize its psychic quality. The overtone series is a law of objective nature as well as of our subjective apperception. The two must not be in conflict.

The second observation concerns the particular intervals that occur in the overtone series. In relation to the fundamental and reduced to the range of one octave, we notice, in this order, the octave, fifth, and major third—in short, the major triad. Then follow the narrow minor seventh, major second (ninth), narrow augmented fourth, narrow major sixth, and major seventh. There are secondary intervals: the fourth between the 3rd and 4th partial tones, the major sixth between the 3rd and 5th, and so forth. The intervals decrease in size with rising pitch; the semitone appears between the 15th and 16th partial tones. The distances above this point are smaller than the smallest interval commonly in use.

The coincidence of the partial tones with those gained by the division of the monochord becomes ever more striking. The results gained by natural and by psychological means are the same. Any number read off the following diagram indicates equally the position of the partial tone and the tone value of the matching division:

Figure 27

At 5, for instance, we find the 5th partial and also the tone reached by the division of the string by 5. Similarly, any intervallic proportion can be read directly from the diagram of the overtone series, for one is contained in the other. Taking only the numerical indications of the overtone series as a starting point, we spot the octave, for instance, at 2:1, 4:2, 6:3, 10:5, et cetera; the fifth at 3:2, 6:4, 15:10, et cetera; the major third at 5:4 and other places; the minor third at 6:5; and so forth. These proportions are identical with the ones developed by division of the monochord. The difference, as emphasized before, is one of approach. The overtones are physical, natural, objective. The intervals are psychological, musical, subjective.

Not only the results agree but the order in which the various tones and intervals make their appearance. The octave, in either case, comes before the fifth; the third, here and there and always, before the seventh. This hierarchy assigns a definite place to each tone in its relation to another tone, whatever the character of each individual tone may be by itself. The prominence of the major triad is evident. The first 6 partial tones contribute exclusively to its formation. Of the first 12 partial tones, only three (7, 9, and 11) do not. It is tempting to refer these facts to the special force of the senarius. By now, however, we should not hesitate to accept the musical experience as a force of equal primacy. *Because* the major triad is a musical event of particular unity, the senarius, besides being a measure, also assumes a formative value.

At the point where the overtone series transgresses the senarius, the flat seventh appears; it has no place in our tone system, nor do any of the subsequent partial tones which occupy the position of a prime number. The artistic utilization of one or the other of these "foreign" tones was handsomely demonstrated by Brahms when he implanted an alpenhorn melody in the finale of his First Symphony (meas. 30 ff.):

10   9 8   6   9   10   8   12   11   9   6   9   10   8

Figure 28

The third observation focuses on the curiously jumpy manner in which the partial tones seem to progress. The hornist cannot possibly play the tones that lie between the fixed points of the overtone series. Whether he wants to or not, he must jump from one to the next. This kind of motion is the opposite of a continuous progression, where the transition from one point to the next is gradual. The overtone series is not continuous. It is formed by "quanta," a word popularized by the relatively recent formulation of the quantum theory in physics. According to this theory, the energy states of atoms and molecules are distributed, not continuously, but in jumps. An embodiment of this revolutionary idea in physics, we submit, exists in the quantum *values* of the overtone series. The tone numbers are not to be degraded to mechanical measures. They appear in the overtone series as audible articulations of discontinuity in nature. Hearing musical entities in the overtone series and calculating energy in the quantum theory are, respectively, subjective and objective apperceptions of the same norm.

Our fourth and last observation derives from the shape of the overtone series, which suggests a progressive contraction. The tones move ever closer to each other. Already after the 16th partial, the intervals are smaller than the musical half-tone; but they continue to shrink into an infinite perspective. Concomitant with this decrease in size is a decrease of loudness. The higher up an overtone, the less audible it becomes. The octave is very loud, and the next few overtones can be easily heard by anybody concentrating on them—as if trying to hear a conversation across the room while a nearer voice threatens to drown it out. The piano tuner checks his work by listening to the overtones, particularly the octave and fifth. The higher overtones become increasingly difficult to isolate. By a curious parallelism, the higher overtones are also much more difficult to produce. A beginner on the horn can soon control the first 8 partials; with practice, he will extend his secure range up to the 12th. But it takes a virtuoso to master a reliable sound of the 16th partial tone, as frequently prescribed by Bach or—to mention a notorious spot—by Beethoven in the scherzo of the "Eroica." The contraction of the intervals and the accompanying decrescendo give the overtone series a definite morphology.

Is there an undertone series? The question has legitimately occupied the fancy of many people. Hugo Riemann, for instance, toward the end of the last century, urgently tried to prove the physical existence of undertones, which he thought he needed to bolster his theory of harmony. It is safe to say that undertones do not exist physically and that future attempts to discover their physical reality are doomed to futility. A string can subdivide itself into smaller parts, but it cannot increase its own length. Hence it can procreate higher, but never lower, tones.

Riemann did not have to prove the physical existence of undertones as a support for the validity of his harmonic thinking. The period in which he lived induced him to accept as "real" only those phenomena that are detectable in nature. To the musician, the physical reality of an undertone series is irrelevant. The psychological reality he does not doubt. He can experience it as a theoretical polarity to the overtone series. He can produce it on the monochord by multiplying a segment of the string (1/8, for instance) in a reciprocal operation to division:

Figure 29

Undertones do not exist spontaneously in nature but they do exist musically in our psyche. The theoretic undertone series thus stipulated permits exactly the same observation that we have connected with the overtone series. The two series mirror each other. They are reciprocal, like left and right hands, in regard to numerical norms, intervals, quantum-like progression, and morphology. They differ, however, like two opposite worlds in regard to their musical tone values. The resulting polarity of the major and minor modes deserves a separate chapter (cf. pp. 189 ff.). At this point, one need only note that the two series together contain all possible tones. Any one tone may belong either exclusively to the overtone series (e.g., g^1 above c) or exclusively to the undertone series (e.g., f_2 below c). The octaves belong jointly to both series (e.g., c^1 above, and c_1 below, c). By developing the two series, one eventually gains any tone.

In view of the observation that the overtone series reaches into infinity, the obvious question arises: how does one determine the pitch of any given overtone, particularly in the range that exceeds our hearing? The answer is easily ascertained when the partial tone in question corresponds in position to a number that can be factorized. What (to pick a random example) is the musical value of the 100th partial of c? We factorize: $100 = 2 \times 2 \times 5 \times 5$; or octave + octave + major third (two octaves up) + major third (two octaves up); or, g^6-sharp. Had we asked, inversely, where in the overtone series of C the tone G-sharp might occur, we would have reasoned thus: G-sharp = major third above major third; or, 5×5; or, 25. This is the earliest appearance of G-sharp, which from here on obviously recurs in every new octave range ($25 \times 2 \times 2 \dots$).

When prime numbers are involved, an extra bit of reasoning will help us determine, or at least approximate, the pitch. We can assume that above 7 all prime numbers will produce pitches that lie outside our tone system. We also know that each prime number above 3 necessarily lies between two products. We further remember from the morphology of the series that each overtone lies slightly closer to its higher neighbor than to its lower. From these premises we deduce the tone value of any prime-numbered partial tone by determining the two pitches between which it lies and by then placing it slightly nearer to the higher pitch. Twenty-nine, for example, is a prime number: what is the pitch of the 29th partial of c? We factorize the numbers of the two surrounding tones. First, $28 = 2 \times 2 \times 7$; or, the flat minor seventh (four octaves up). Then, $30 = 2 \times 3 \times 5$; or, the major seventh (four octaves up). The 29th partial of c, accordingly, lies four octaves up between a low b^4-flat and b^4-natural, slightly closer to the latter. (To call this tone a high B-flat or a low B-natural is irrelevant; because it does not exist in our system, it has no appropriate name.) Any such tone may be made to sound on the monochord. The 29th partial, for instance, lies at 1/29 of the total string length, that is, at 120/29 cm, or at 4.137 cm, on our monochord of 120 cm. Successive multiplications by 2 will bring this pitch down to the desired octave range (e.g., at 16/29 or 66.192 cm in the first octave).

Exercises

1. What are the 14th, 16th, 18th, 19th, 20th, 30th, 37th, 43rd, 50th, and 800th overtones of *c*? You may express the pitches by applying octave reduction.

2. Sound all these pitches on the monochord. At which centimeter does each lie when reduced to the first octave above *c*?

3. Exactly where in the overtone or undertone series of *c* do the following tones of our tonal system occur: *c*-sharp, *d*-flat, *d*, *d*-sharp, *e*-flat, *e*, *f*, *f*-sharp, *g*-flat, *g*, *g*-sharp, *a*-flat, *a*, *a*-sharp, *b*-flat, *b*?

4. Set up a siren, according to the instructions outlined below. A siren is the perfect instrument, not only for demonstrating a total glissando, but for helping us gain an insight into the correspondence of fixed-tone relations and geometry.

The following equipment is needed: (a) A small electric motor. (b) A source of compressed air. Bottled air, or an air pump, or a reversible vacuum cleaner will do, provided the installation does not produce any disturbing noise. (c) One or several disks of stiff material, such as metal or masonite, about one foot in diameter. Describe several concentric circles on the disk; and on each of the circles construct a regular polygon. If you have, for instance, five circles, you may construct an equilateral triangle on one, a square on the second, a pentagon on the next, a hexagon on the following, and a heptagon on the last. On the apex of every angle on every polygon, bore holes with a quarter-inch drill. (The most advantageous size of the hole must be found experimentally, for it depends on the air pressure and the nozzle size you will be using.) Drill a hole in the center of the disk, or disks, as the case may be, with the diameter corresponding to that of the sleeve of the motor shaft.

Now connect your source of compressed air to a nozzle by means of a flexible tube. Mount a disk on the sleeve of the motor shaft. Turn on the air, start the motor, and hold the nozzle near the rotating disk opposite one of the concentric circles. A tone is heard, of which the frequency is the product of the number of holes and the number of revolutions per second. For example, the square rotating at 1725 r.p.m. produces a pitch with the frequency of $4 \times 1725/60 = 115$. The corresponding tone can be determined by our reasoning that 115 lies near $a_2 = 110$ (i.e., 440/4). The smallest semitone above this pitch produces b_2-flat $= 116.3$ (i.e., $110 \times 16/15$). A frequency of 115 accordingly corresponds to a somewhat low b_2-flat.

If we now hold the nozzle successively opposite the various circles and polygons, we shall hear tones that form the following intervals with the tone produced by the square:

Triangle: fourth below (4/3).

Pentagon: major third above (4/5).

Hexagon: fifth above (4/6 = 2/3).

Heptagon: natural minor seventh above (4/7).

If we let the motor die down gradually while the air is blowing against a disk, a glissando down to zero will be the audible result.

The symbolism of regular polygons, so important throughout antiquity and the Middle Ages, is here experienced directly as a musically significant event. The people of former times were well aware of this connection. St. Ambrose says that the octagonal shape of baptisteries and baptismal fonts alludes to the octave, "which renews the whole man."[1]

---

[1] *Epist. class.* i. 44. (*Patrol.* xvi. 1140.)

4 | Tone Properties

GENERAL CHARACTERISTICS

A tone has three variables: pitch, loudness, and timbre. We say of a tone that it is high or low; loud or soft; and characteristic of one kind of voice or instrument or another. We could add that a tone also has a certain duration; but this quality, though of essential importance in music, does not pertain to tone alone but is shared by all temporal experiences.

For a good musical reason, pitch has occupied our attention initially more than the other tone characteristics and will continue to do so. To the physicist, all properties of a tone are rightly of equal interest. To the musician, however, a discrimination between structural and accessory tone characteristics is meaningful and essential. We can easily think of a musical composition in terms of pitch relations (and rhythm, which does not concern us here) to the complete exclusion of loudness and timbre. Bach's *The Art of Fugue* is only an obvious, and by no means a unique, example that proves the point. On the other hand, the musician cannot possibly imagine a given composition entirely in terms of loudness and timbre without reference to pitch. Test these assertions by thinking of a Mozart sonata or Bach chorale you know. Jointly with the rhythmic structure, it is pitch that identifies the work. Pieces individualized only by loudness and timbre (as, e.g., an ensemble for percussion instruments without definite pitch) have been written experimentally; but except for limited use within a larger framework, this kind of writing makes a minimal contribution to music.

PITCH

Experiments on the monochord have helped us develop the basic laws of pitch relationships. A shorter string agrees with a higher pitch, that is, a higher frequency. String length and frequency are reciprocal.

Because not even the lowest musical frequencies can be counted by the eye or by any simple method, rather complicated apparatuses are necessary to effect an exact measurement. An oscillometer, as found in most physics laboratories, measures frequency with the help of electrical impulses. An oscillograph translates these impulses into a visual record. The musician really does not depend on these, and similar, machines. The proportions gained on the monochord suffice. If the musician knows that, by agreement, the tone produced by 440 vibrations per second is called a, then the octave above, a^1, has a frequency of 880; the octave below, a_1, a frequency of 220; the fifth above, e^1, a frequency of 660. In short, he can easily compute any frequency by applying to a given fixed frequency the inverted fraction found on the monochord.

Is the identification of a with the frequency of 440 (i.e., of middle c with 264) really just an agreement? It seems so to us, although there is some evidence to the contrary.

In Western tradition, the powers of 2 have served as points of orientation; the practice of doubling physical units is a good example of this principle. The organ builder, combining the theoretical and practical aspects of music, sets the lowest usable subcontra c_4 at the lowest humanly audible frequency of 16 $(= 2^4)$. Middle c accordingly appears as 2^8 or at a frequency of 256; and a, a sixth above, would thus correspond to $256 \times 5/3$, or 426.6. This absolute pitch may sound a bit low to us but it has been defended by various musicians from old to modern days as more "natural" than our present convention.

The Chinese reached the almost identical result by a totally different method. In traditional Chinese music, the pitch norm was a tone, called *huang chung* ('yellow bell'), which sounds like our f-sharp and corresponds, as subsequent measurement revealed, to a frequency of 366. This tone and the pipe measures that defined it were derived from certain sacred cosmic proportions, which recur throughout Chinese philosophy and religion. According to the standard set by *huang chung,* the frequency of *a* would be almost exactly 440 (a minor third above *f*-sharp: $366 \times 6/5 = 439.2$).

The search for an objective pitch measure is old and universal. The Occidental took the measure from man; the Oriental, from the cosmos. The two measures for absolute pitch, arrived at from opposite directions, coincide almost exactly.

In line with the modern tendency to standardize, an international conference in London in 1939 agreed that *a* should equal 440 Hz. To secure the standard, tuning forks are built accordingly. Radio stations in many countries, particularly on short wave, regularly broadcast the standardized pitch for orientation and guidance. In the United States, it can be heard at stated periods throughout the day over the stations WWV and WWVH of the National Bureau of Standards. In Switzerland, one may dial a certain telephone number to receive the standardized *a*. The dictatorial regime in Italy before the Second World War issued detailed decrees that pinned the responsibility for tuning to, and playing at, the standardized pitch on particular performers (the conductor, the first violinist, the manager, as the case may be) and that threatened to punish any unwarranted deviation by money fines and jail sentences.

The maintenance of a standardized pitch has always been difficult and problematic. In a misdirected effort toward greater brilliancy, most American orchestras today tune to an *a* of which the frequency lies above 440. The pitch of the Boston Symphony Orchestra, among the highest in the land, is oriented to an *a* reportedly at 446. The opera orchestra in Vienna usually plays at an equally high pitch, to the distress of visiting singers who risk sounding flat unless they strain and readjust. The natural structure of a singer's larynx and vocal cords, the natural supremacy of the voice over instruments (cf. pp. 69 ff.), provides the safeguard against an uncontrolled raising of the absolute pitch. If Beethoven's choral parts, for instance, sound fearfully high today, the error does not rest with the composer: the many tenor *a*'s heard today were *a*-flats in his day. Stringed instruments can be easily tuned higher by a turn of the peg. Wind instruments have been built ever shorter to conform to the rising pitch. The human voice, here as elsewhere, reminds us of the existence of natural norms, which must not be transgressed indiscriminately.

The London agreement of 1939 was not the first of its kind. A French Government Commission in 1858 set a at 435 vibrations per second, and a special international conference on pitch in Vienna in 1885 gave the official sanctioning to this standard. Yet orchestras in England at the time tuned to a lying anywhere between 438 and 455, and the Leipzig Gewandhaus concerts were measured at $a = 448$ in 1869. The discrepancies were even greater before the French recommendation, although musicians, while often fighting over the absolute pitch, seldom suffered from the lack of uniformity. They were safe in their knowledge of proportions, which had universal validity. They were free of the bondage of absolute measurement, primarily because each local community and church could consider itself self-sufficient. When Arp Schnitger installed an organ in St. Jacobi in Hamburg in 1688, he tuned it to $a = 489$; whereas Andreas Silbermann in the Strasbourg Cathedral twenty-five years later took his orientation from $a = 393$. A famous quarrel occurred in Cremona in 1583 when the municipal organ repairman, Gian Francesco Mainero, who had been commissioned to replace deficient pipes, decided on his own to lower the pitch of the whole instrument by one halftone. The church organist Marc'Antonio Ingegneri prevailed at committee meetings and public hearings to keep the pitch high, on the grounds that the instrument would otherwise lose sonority and "vivacity of spirit." Handel's and Mozart's tuning forks, which are preserved, yield an a of 422; and this was probably also Bach's pitch. Actually, Bach operated on at least two different standards, the chamber pitch and the choir pitch. The latter was about one whole-tone higher than the former. For this reason—to give only one of many examples—the original organ parts for the *Passion According to St. Matthew* are written in D minor, a major second lower than the tonic key of E minor; the higher choir pitch of the organ accomplished the transposition. In any case, we should never forget in the heat of arguments on the aesthetic character of keys that the symphonies of the Vienna Classic Masters sounded a half-tone lower to the composers than they do to us today.

How did musicians measure and designate the absolute pitch of a tone before the invention of modern devices? It is reported that the Chinese, around the year 1500 B.C., but probably even earlier, filled the pipes of a wind instrument with small, hard, bean-like kernels and then counted the kernels. Assuming that the kernels were all more or less of the same size, and the pipes of the same diameter, one could then specify any pitch by the number of kernels needed to fill a pipe. Everything else being equal, two pipes containing 40 and 80 kernels, respectively, would be an octave apart, the larger number representing the lower tone. A pitch pipe called "lü" set the absolute standard. Organ builders throughout the ages have taken the pipe length of a normally pitched flue-stop, such as a principal, or open diapason, as the measuring unit. The note c_2 has been identified with a pipe length of 8 feet. All other designations derive from this reference point. The pipe c_3, one octave lower, is a "sixteen-foot" (usually written as 16'); c_1, one octave higher, a "four-foot" (4'). The tone $a = 440$, in the jargon of organists, is the a in the two-foot octave, or, quite precisely, the $1\frac{1}{5}$' (i.e., $2' \times 3/5$).

The length of an organ pipe, the number of bean kernels, the string on the monochord: they all are spatial materializations of the acoustical experience. The concept of space is inextricably connected with the idea of pitch. Higher tones presuppose smaller pipes, a smaller number of beans, shorter strings. Deeper tones presuppose larger vibrating media. There is a decisive qualitative difference between the identification of a pitch by a frequency or by a wavelength, the mathematical equivalence notwithstanding. Frequency is an acoustical measurement of time; wavelength, of space. The temporal character of frequency is easily coordinated with rhythm. We think of a string, or of an air wave in a pipe, or of any tone-producing body as moving to and fro within a given time unit. The frequency of the vibration phenomenalizes time; for the successive frequencies of any tone maintain the same temporal distances, in short, the same rhythm. The spatial character of wavelength, on the other hand, impresses us primarily not as a rhythmic motion but as a perceptible shape. We visualize the length of a string, or the space of an air column, or the weight and size of a bell. The wavelength gives form to space.

The time-space question has occupied the minds of philosophers throughout the ages. It is stated at the beginning of the Bible when God creates three-dimensional space by identifying heaven and earth, and then creates time by *saying*, "Let there be light"—an acoustical event that leads to the rhythm of the first day. We content ourselves here with the recognition that wavelength and frequency represent to the musician the reciprocity of space and time, of being and becoming. Tone exists in space as a wavelength, and moves in time as a frequency. Shown proportionately, this reciprocity of space and time appears in the following graph equidistant on one side, and perspectively contracting on the other:

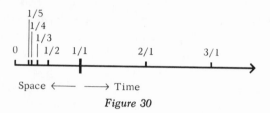

Space ⟵—— ——⟶ Time

Figure 30

Exercises

1. What are the frequencies of the lowest and highest tones of your piano (a = 440)?
2. What are the frequencies of the four open violin strings?
3. What are the wavelengths of the four open violin strings?
4. What are the frequencies of the four open violoncello strings?
5. Compute the frequencies for all tones of the C-major scale from c to c^1.

LOUDNESS

Experiments on the monochord soon persuade us that the loudness of a tone is related to the degree to which the string is disturbed. The degree of the disturbance can be measured by the perpendicular distance between the line formed by the position of rest and the farthest point reached by the vibrating string. This distance is called the "amplitude" of the vibration. Loudness depends exclusively on amplitude.

The performing musician who keeps this fact in mind will be protected against illusory notions regarding his control of loudness. We shall discuss detailed applications in the chapters on tone producers (pp. 69-162).

When the loudness of a tone fluctuates, one speaks of a "tremolo." This term should be carefully separated from the fluctuation of the pitch of a tone, which is called "vibrato" when the pitches remain undefined, and "trill" when the pitches become discrete.

Here the tyro sometimes puzzles over the speed at which the string oscillates back and forth and which seems to be somehow connected with the amplitude and even with the frequency. The best safeguard against any possible confusion is to ignore altogether the concept of speed of a vibration and to adhere to the clearly defined concepts of frequency (or wavelength) and amplitude. The speed of the vibration is merely a by-product. Let us assume that a string corresponding to $a = 440$ is plucked, first, to produce a soft tone, and, later, to produce a loud tone. In the first instance, the amplitude, the "belly" of the vibration, is small; in the latter, large. In each case, however, the string completes 440 vibrations per second. Tuned to a certain pitch, in this case to a, it has no choice but to swing back and forth 440 times per second, regardless of the loudness. Now it stands to reason that the string covering the wider distance in the same time unit has to move faster, but the frequency is not affected. If two boys are asked to run to the end of their block and back in the same time unit but if one boy's block is longer than the other, this boy will have to run faster. The two boys complete the round trip in the same time (i. e., the frequency is the same for both), but the boy covering the wider distance has to run at a higher speed (i.e., his amplitude, his "loudness," is greater).

One should distinguish the measurable amount of energy in the sound source, the intensity, from the impression of loudness in the subject. The relation between an increase in intensity and the corresponding increase in loudness has practical implications. Loudness does not increase in direct proportion to the source energy but rather follows the Weber-Fechner law. This law states that, in general, the response of any of our senses increases approximately as the logarithm of the stimulus. A common illustration of this law is provided by a three-way light bulb: our eyes perceive the difference from 100 watts to 150 watts as smaller than that from 50 watts to 100 watts. Our aural experience, analogously, tells us that above a certain increase in the number of identical instruments, the return in loudness tends to become negligible. The nine trumpets playing in unison at the beginning of Janáček's *Sinfonietta* add to the loudness of the phrase less impressively than their appearance might promise.

The following table shows, in successive columns, the number of identical instruments (e.g., violins) used; the corresponding logarithms; the difference between adjacent logarithms; the approximate percentage of the increase in loudness between two adjacent numbers of instruments; and the approximate percentage of the increase in loudness over one instrument.

Number of Violins	Logarithm	Logarithmic Difference	Loudness Gain between Adjacent Numbers	Loudness Gain Over One Violin
1	0.0000	0.0000	0%	0%
2	0.3010	0.3010	30%	30%
3	0.4771	0.1740	17%	47%
4	0.6020	0.1249	12%	59%
5	0.6989	0.0969	10%	69%
6	0.7781	0.0792	8%	77%
7	0.8451	0.0670	7%	84%
8	0.9030	0.0579	6%	90%
9	0.9542	0.0512	5%	95%
10	1.0000	0.0458	5%	100%

A loudness gain of 100%, that is, a doubling of the initial loudness, is accomplished by increasing the number of violins from 1 to 10 (or from 2 to 20, from 10 to 100, etc.). If one merely doubled the number of violins (from 1 to 2, from 2 to 4, from 3 to 6, etc.), the loudness would increase only about 30%. Although the Weber-Fechner law is valid only as an approximation, it proves sufficiently that the difference between, say, ten and twenty violins is "uneconomical" in the production of loudness. Yet something is to be gained from such an increase. Twenty violins playing the same tone have a smoothness and richness of timbre that can hardly be obtained by ten. Herein lies the real value of a large number of violins in an orchestra, or of singers in a chorus.

Exercises

1. Make a plumb line and use it as a pendulum. Change the amplitude repeatedly, and in each case count the number of oscillations within a time unit. As long as extremely small and extremely large amplitudes are avoided,

notice that the frequency remains constant. In order to change the frequency, alter the length of the pendulum.

2. Assemble a number of friends. Ask each of them to sing a tone of a certain pitch and at a certain loudness. Then have two and more of them sing the tone and observe how slowly the loudness increases.

3. Try the same experiment with tones of different pitches and different degrees of loudness.

4. While attending an orchestra performance, note how easily the sound from an instrument of moderate strength is drowned out when embedded in sounds from other instruments in the same pitch range; and how, on the contrary, a relatively soft instrument playing in a different pitch range can be prominent (e.g., the flute).

TIMBRE

When a violin and oboe, for instance, play the same pitch with the same loudness, we can yet distinguish the two tones by their timbre. This property is often called the "tone quality" or the "tone color." We advocate ignoring these names, because they tend to create confusion. Tone "quality" is too general, for logically pitch and loudness are also qualities. Tone "color" is an optical term which is out of place in acoustics.

Timbre is easily distinguished even by untrained ears. The physical counterpart is based on the natural fact according to which a tone is not just "one thing" but a multitude of combined partial tones. Theoretically, the overtone series as a whole is always the same in relation to a fundamental. Practically, the overtone series is always modified by the nature of the tone producer. The construction of an instrument favors the loudness of some overtones at the expense of others; it extends the overtone series far up or limits it close down; and it may even eliminate some overtones altogether. A different constellation of overtones is responsible for a different shape, or form, or complexity of the vibration. The differences between overtone constellations account for the differences of timbre.

This generally accepted explanation of timbre distinctions between single tones has several corollaries. One, for instance, concerns the indirect bearing of loudness on timbre. For as the loudness of a tone increases, overtones are aroused that may have been negligible before; and as it wanes, some overtones are dampened and lost before others. The result is a change of timbre concomitant with a change of loudness.

Another significant contribution to the total timbre formation is made by the manner in which a tone begins and ends. The attack, in particular, influences the characteristic overtone constellation incisively. Although the initial transient sounds quickly disappear as the vibration producing a tone becomes stabilized, they seem to possess such an identifying force that a comparison of two different timbres without them often becomes most perplexing. Here lies one reason for the difference—even on the same instrument—between the "beautiful tone" of an artist and the less satisfactory product of a beginner.

The particular instrument bears the greatest responsibility for the potential timbre. The resonances of two violins, for instance, will necessarily be not identical; and the resulting timbres, accordingly, differentiated. What makes a violin, nevertheless, always sound like a violin although the various tones on the instrument are likely to show variants among the respective overtone constellations? An answer is supplied by the "formant theory," according to which the partials of a certain range on any instrument are characteristically favored regardless of the ever-shifting fundamentals. If we have understood the timbre of a single tone to be formed by the relative distribution of its overtones, we may say that the predominant timbre of an instrument is formed by the partials that lie within an absolute pitch range, the "formant."

The variety of timbre is unlimited, because overtones run theoretically into infinity, and so do the possible combinations of the occurring pitches and degrees of loudness. Moreover, each overtone begets an overtone series of its own and thus adds branch to branch. A particular timbre reflects a particular overtone constellation. A "pure" tone, that is, one free of overtones, can be artificially produced in a laboratory, although here, too, some overtones are likely to be formed by concomitant circumstances. A tuning fork, to be practical and neutral, is as free of overtones as possible. The expressive power of a tuning fork is accordingly insignificant.

The control of timbre by the musician—be he the composer, performer, or instrument builder—will be discussed at length in the relevant chapters on the human voice and music instruments. At this point, the identification of timbre with the shape of the vibration, that is, with the particular overtone constellation, may suffice.

Exercises

1. Depress the right pedal of a piano and play a loud tone in the middle or bass register. Keep the key depressed. After a while, release the pedal and hear the marked change in timbre. The tone loses richness because all those strings are now dampened of which the resonance had initially reinforced the overtones of the struck string.

2. Pluck the monochord string at different points. (Bowing or striking yields comparable but less immediately evident results.) Try especially 1/2, 1/3, 1/5, and 1/7. You will hear a variety of timbres, because all partials with a node at the point of contact will be missing.

3. Cut two cardboard mailing tubes of different diameters to the same length. Blow across each and listen to the difference in timbre.

INTERFERENCE

In almost any practical musical situation, more than one tone will be in play at any one moment. The effect of two (or more) tones on each other should be understood by the musician. The term "interference" logically comprises all phenomena eventuated by the influence of several wave trains on one another, although in colloquial usage it denotes primarily the disturbance of a process. The results of interference, in the widest sense, vary with the relationship of the frequencies involved.

Let us first assume two sound waves of the same frequency. The state of a wave at a certain point in space and at a certain moment in time is called "phase." Two waves, at a given point and at the same moment, may have coinciding or conflicting phases. If the vibrations are "in phase," they reinforce each other. If they are "out of phase," greater or lesser cancellation occurs. If they are "in opposite phases," the sound is totally annihilated. Think of a person pushing a child on a swing. If he moves in phase with the swing, some of his energy is added to that of the oscillating swing. As a result, the swing flies higher. The amplitude is increased. In musical terms, the tone is louder. Two violins, or sixteen violins in an orchestra, playing in unison produce a louder tone than one violin. If the same person now pushes out of phase with the swing, he interferes with the motion of the swing so as to reduce the amplitude. If he is in opposite phase so that he hits the swing as it moves toward him, the whole oscillation will ultimately come to a halt. In the comparable musical situation, imagine two violinists playing in unison (although we concede that the probability of an exact unison is about zero). Depending on the relative phase at any one point of the auditorium, the tones may reinforce, or weaken, or cancel each other at that point. Because two, let alone sixteen, violins are not likely to play in a mathematically perfect unison, a listener at any one point of the auditorium will hear a rapid succession of all three possibilities. The average result is basically an increase in loudness with each additional violin. Moreover, the successive irregularities in the amplitude will produce an undulation of loudness, which the musician anywhere in the hall recognizes as a "tremolo." The situation involves greater risks if two identical organ pipes were placed next to each other with their mouths facing each other. They may become a coupled system of such a kind that the tones would actually annihilate each other. Hence every organ builder avoids this placement.

If the sound waves have different frequencies, they all contribute toward an increasingly complex form of the vibration of the air. Each new frequency adds its characteristic to the curve of the vibration. There is no spatial limit to the possible modifications. In practical terms, the different frequencies sent out by an orchestra, for instance, combine into one complex vibration of the air; but the degree of complexity never becomes saturated and is not limited by such factors as the size of the hall or the nature of the air. When the complex sound wave reaches the hearer, the ear dissolves it into its components in a process not elucidated, but scientifically paralleled, by Fourier's method of sound analysis (cf. p. 10).

Two special cases of interference have important practical implications for the musician. In one, the frequencies of two tones express a simple proportion and produce combination tones. In the other, the frequencies of two tones lie very near each other and produce beats.

The famous eighteenth-century violinist Giuseppe Tartini is credited with first discovering, that is, hearing, a lower, third tone, a "terzo suono," while playing two fairly loud other tones simultaneously. The frequency of this third tone is the difference of the two primary frequencies; and the name, accordingly, "difference tone." A third tone that lies higher than the two generating tones also exists but is less easily heard. The frequency of this third tone is the sum of the two primary frequencies; and the name, accordingly, "summation tone." Collectively, we refer to difference and summation tones as "combination tones." We need not be detained by the old argument whether the combination tones exist "objectively" or only "subjectively." We have defined tone as something that, in any case, happens within us, and we therefore do not doubt the musical reality of combination tones. Summation tones are of little importance in music, because they are difficult to hear. Difference tones, on the other hand, are commonly utilized in police whistles, which yield a comparatively low tone with the help of two very short pipes of somewhat unequal length. Difference tones also enable the organ builder to produce a very low pitch without the practical difficulty of having to supply a very big pipe. The best combination tones are the product of two primary tones of which the frequencies express a simple senaric proportion.

Here are a few examples of difference tones:

Figure 31

If two tones of nearly the same frequency sound together, we hear a rapid waxing and waning in loudness. This phenomenon is known as the "beats" between two tones. The occurrence is due to the shifting of phases between the wave trains; the result is a periodic addition and subtraction of amplitudes. The number of beats is always equal to the difference between the two frequencies. Accordingly, the same small interval has more beats in a higher than in a lower octave range. Beats provide a check on intonation, for the undulation of loudness becomes less rapid as the frequencies of two slightly discrepant strings are made to approach each other. Two A-strings, for example, tuned to 440 and 444 produce four beats per second, which are gradually reduced to zero as the higher string is tuned down to the perfect unison.

Exercises

1. While slowly rotating a vibrating tuning fork, hold it now nearer to one ear, now at an equal distance from both ears. Observe the change in loudness, and note the corresponding position of the prongs.

2. At a position of least loudness, insert a cardboard between the prongs. Because the cardboard eliminates interference, the tone will suddenly be louder.

3. To hear beats easily, play a minor second in a very low register on an organ or harmonium.

5 | The Human Voice

ORGANIC CONSIDERATIONS

Music manifests itself through man and in man. Goethe said: "It issues in the voice and returns through the ear, stimulating the whole body to join, and evoking an inspiration of feeling and behavior and a cultivation of the inner and outer senses." The organic musical function is divided between the voice and the ear. The productivity of the voice is awakened and stimulated by the transformative powers of the ear; but both voice and ear are necessary for the total subjective musical experience. The eye, by comparison, both registers and expresses, receives and conveys. The ear, as Goethe amusingly puts it, is a mute sense. The voice supplies the missing quality.

The voice is the primary subjective tone producer, compared to which all music instruments are representatives of the outside objective world. Singing is a totally productive activity, inasmuch as, like composing, it directly creates music "from the inside out." Thus the merit of singing depends, not only on the physiological nature of a particular throat, but even more so on the mental guidance by one's trained ear.

Two important conclusions arise from this fact—one for the singer in particular, the other for the musician in general.

More than an instrumentalist, the singer has to be guided by a clear and precise preconception of the sound he wishes to produce. He might profitably learn to know how the muscles involved in singing operate in order to respect the natural limitations imposed on his art. But no knowledge will force the vocal muscles into obedience unless the musical ear first dictates the precise demands.

The second conclusion, for the musician in general, stems from the recognition that the human voice, because it exists as a naturally given melodic tool, sets up melodic norms which composers must not ignore and which listeners can spontaneously understand. In the specific sections below dealing with the vocal control of pitch, loudness, and timbre, the relevant emerging norms will be developed. At this initial point, one need only consider the standard set by the length of the human breath. We judge a melody as "long" if it stretches beyond the exhaustion of one breath; and as "short" if it does not expend it fully. Most fugue themes by Bach, for instance, can be sung on one breath. If they require more time, as, for example, in the first Kyrie of the B-minor Mass or in the big organ fugues in D major and G minor, they form, not a single melodic unit, but a strongly partitioned whole that binds smaller melodic phrases into an entity. We accept as a unit a melody that can be rendered on one breath.

How elemental the natural force of singing is, the close connection of the human voice with sex amply evinces. The polarity of male and female is audible in the voice distinctions. Castration, as practiced in the eighteenth century, produced an artificial female voice in boys; and the evil of perversity was matched by the reward of rarity. The changing of a childish to a mature voice occurs as a symptom of puberty. At the other end of the life curve, the voice wanes together with the sexual powers. A singer's career coincides with the period of reproductive maturity. The direct sexual function of singing is even more obvious in animals such as birds, where the male warbles to woo the female.

The human vocal apparatus has been studied and described by biologists and physicists. The results of detailed technical investigations may be looked up in any of a number of texts on physiology. Here, we may briefly state the basic facts in order to devote our main attention to the musically relevant concepts.

The actual tone producer, that is, the body that sets up the primary vibration, is a pair of muscles in the larynx called the "vocal cords." The vibration is started by an air stream sent against them from the lungs. The vocal cords lie next to each other in such a way that air may pass between them or be blocked. The opening between them is called the "glottis." The vocal cords can be stretched and relaxed in varying degrees. They can open and close the glottis in varying degrees. A singer controls the action of the muscles in his throat no more consciously than a runner, for instance, the action of the muscles in his legs. The singer or the athlete imbues the potential of a muscle with purposeful life by a clear idea of what he wants to do. All he need know about the muscle is how to protect it against damage by not making excessive demands of it.

In addition to the source of air supplied by the lungs and the vibration produced by the vocal cords, at least two other physiological agents participate in the formation of the human voice. The tone is amplified, and it is articulated. Amplification, here as on the monochord, results from the free and forced resonances of the surrounding parts. Bones and cavities throughout the head and the upper part of the body participate. Articulation of the sound, finally, is accomplished by the activity of the mouth and all its components.

The manner in which these four agents—wind supply, tone producer, amplifiers, and articulators—cooperate in producing a tone of the human voice, the particular sections on pitch, loudness, and timbre, which follow, will attempt to elucidate.

PITCH

The tonal space covered by the totality of the human voices is evenly distributed between the two sexes. This manifestation of a fundamental polarity permeates the entire field; for the male and female ranges each are further divided into a high and a low voice—bass and tenor correspond, respectively, to alto and soprano which duplicate them at the octave. The principle of polarity is further evidenced by the movement of any voice which is either upward or downward.

An untrained voice spans about one and one-half octaves. The four voice ranges overlap, middle *c* being common to all of them. The total average range of the combined human voices covers about three octaves:

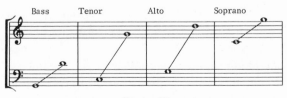

Figure 32

This average situation is expanded by trained voices to almost four octaves:

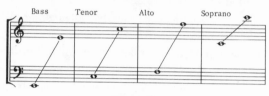

Figure 33

Physiologically, we can hear tones beyond the extremes of the human voice at either end. Music instruments make ample use of both higher and lower tones. The piccolo, the highest instrument in the orchestra, surpasses the high c^2 of the soprano by two octaves; and the double bassoon underprops the human bass by well more than an octave. Yet the range, like all the other qualities of the human voice, sets a norm that influences all our musical thinking and intelligibly regulates any musical deviations. Thus we accept as the central musical "playground" precisely that area which all four vocal ranges share —the short span around middle *c*. Most of our traditional musical works, even when played by instruments, are framed by the total vocal range. The early keyboards kept more or less to the four human octaves. Mozart never went above f^2 on the piano, the top accomplishment also of the Queen of the Night in *Die Zauberflöte*. The traditional symphony orchestra moves primarily in the circumscribed area, transgressing in either direction mostly by octave duplications.

Each voice, although forming part of the total span, must also be considered as a whole by itself. Each has its own bottom, middle, and top. Besides being related to the other human ranges, each voice thus is heard and understood as an aesthetic entity.

The singing of tones in extreme ranges, like any extreme action, is easily recognized as a special accomplishment. If the soprano nevertheless outshines the alto, and the tenor the bass, the physiological apparatus of the human voice justifies the verdict.

The pitch of the human voice is controlled by the vocal cords. The singer— one cannot emphasize this fact too often—controls the vocal cords by his musical imagination rather than by muscular consciousness. As the pitch rises, the vocal cords gradually stretch and grow thinner; for very high pitches, they are furthermore shortened. The glottis, in consequence, becomes ever smaller. For lowering the pitch, the vocal cords gradually relax and grow thicker, with a concomitant opening of the glottis. Each cord thus functions similarly to a vibrating string, whereas the two cords jointly act like a double reed. Our psychology, for good reasons, considers tension a deliberate accomplishment as compared to relaxation. The acrobat tightening his muscles is admired rather than the person reposing in an armchair. Similarily, the high *C* is understood (and recompensed) as a rousing feat—much more so than its counterpart at the other end of the scale, which is not easily produced, either. Sopranos and tenors in general carry away the glitter attributable to stars. The case is rare when a bass conquers the limelight because of his particularly low tones. In Verdi's *Rigoletto*, Sparafucile's low f_2 is no real competition for the Duke's high *b*.

The human voice is the primary producer of melody. It therefore sets natural norms for the concept of melody. The pitch mechanism of the vocal cords, apart from making us understand questions of range, also establishes a criterion for melodic cohesion. The vocal cords necessarily stretch and relax gradually and not by jumps. To change from one pitch to another, they therefore must pass through all intermediate stages. Accordingly we recognize conjunct motion as a basic melodic principle. Progression by step is the norm. A skip is a deliberate effort to overcome the bondage, and it is properly heard as a melodic accomplishment of a special kind. Gregorian melodies embody this principle in its purest form. Certain melodic types of recent centuries, on the other hand, for example, the opening of the "Eroica" Symphony, stem so strongly from harmonic sources that they have been appropriately called an "unfolding of a harmony in time" rather than a primary melody.

Quite literally, the human voice can produce connected tones, that is, a legato line, only in a glissando. A good singer can give the impression of singing a skip legato by mastering two technical devices: he minimizes the time lapse between the two tones of the skip, and he reduces the breath he lets out during this time lapse. In this sense, the legato singing of any skip involves a portamento, which the good singer can render almost inaudible (if he wishes) by a simultaneous split-second reduction of loudness. This technique is particularly efficient and easy when the discrete tones of a melody lie adjoining to each other.

At this point it is well to reiterate that a good singer produces a pitch, not by working his vocal muscles in a deliberate manner, but rather by concentrating on a clear mental concept of the desired tone.

LOUDNESS

Any person can produce loud and soft vocal sounds at will. Even the totally untrained and even unmusical man, who cannot match the pitch of a tone or hear a difference in timbre, will readily control dynamic differences in his own, and recognize them in somebody else's, voice. Yet no aspect of the tone produced by the human voice is less clearly understood by the experts who have investigated it. The best among them confess to complexities that have not yet been properly explained. Their hope for more elucidating research in the future is shared by us.

A few basic facts emerge. Obviously the amplitude of the vibrating vocal cords is directly connected with the loudness of the resulting tone; but an excessive amplitude, apart from damaging the muscles, might equally damage the intonation. Think of a pendulum in an old grandfather's clock. Within a certain amplitude, the time needed for one oscillation, that is, the frequency, remains constant, regardless of whether the actual path traversed is longer or shorter. Beyond a critical amplitude, however, the longer path stretches the time the pendulum needs to swing from one extreme point to the other; and the frequency changes somewhat with the amplitude. Precision clocks in some scientific institutes use pendulums of enormous length (over thirty feet in the Vienna Urania, for instance); the frequency is thus protected against minor irregularities caused by the fluctuation of external mechanical forces.

A singer, similarly, can begin to suffer from faulty intonation if his control of loudness relies exclusively on the action of the vibrating vocal cords. Actually, loudness of the human voice depends on at least two factors and on the intricate and ever varying ratio between them: air pressure from within, and resistance of the vocal cords against this pressure. The amplitude of the vibration is the resultant of these forces.

The physiological process by which diaphragm, abdomen, ribs, and a multitude of muscles all cooperate to press air out of the lungs does not concern us here. The musician is content to remember that pressure will influence both the quantity of air that passes through the glottis at any given moment and the speed with which this quantity escapes. To get more water out of a faucet, one must either increase the pressure (which the user in an apartment normally cannot do) or widen the opening (which a turn of the faucet accomplishes). The singer has both pressure and opening under his control at all times. To master the loudness of his voice, that is, the amplitude of the desired vibration, he must constantly adjust one against the other. Too much air pressure would remain inaudible if there were no resistance of an elastic body (in this case, the vocal cords) against it to translate the flow of air into vibration. Too much resistance of the vocal cords, however, might block the flow of air altogether. The optimum opening or closing of the glottis is more easily heard than scientifically defined. Clearly, in a louder tone as produced by more air pressure, the growing amplitude of the vibration of the vocal cords also increases the amount of air that may pass through the glottis.

The ratio between these two kinds of tensions—air pressure and glottis resistance—determines the quality, not only of an individual singer's art, but also of temporally conditioned styles of singing. The old *bel canto* technique, for instance, reduces the air consumption to a minimum and consequently demands an optimum glottis closure. The resulting tone is relatively small but all the more flexible. The large halls of the present period, on the other hand, exact a large tone, for which high air pressure and ample breath consumption have become prerequisites. The glottis resistance is accordingly more rigid; and an atrophy of the florid style of singing has been the inevitable price paid for the dynamic gain.

The coordination of the two forces that control the loudness of the human voice, that is, the amplitude of the vibration of the vocal cords, requires a correct ratio for every tone sung. The technical problem would seem insurmountable if a natural coordination, here as in many other human activities, did not contribute to a solution. This is particularly true in a "normal" situation—that is, when a tone is sung neither too loudly nor too softly in the middle range of a particular voice. The difficulties of coordination increase as this middle norm is abandoned for volume control in extreme ranges. Very few tenors can actually sing the closing high b-flat at the end of *Celeste Aida* as pianissimo as prescribed (and obviously expected) by Verdi. Toscanini once asked an otherwise excellent Radames quickly to abandon the high b-flat, which came out forte, and to end the aria one octave lower on some inserted words in order to salvage the pianissimo ending at all cost.

Without ever minimizing the principal forces responsible for vocal dynamics, one must not forget that resonances play an audible part. As various cavities in head and chest vibrate in free or forced resonance along with the vocal cords, the loudness of a tone undergoes decisive modifications. These dynamic enforcements are important but secondary. Their main significance applies to the timbre of a tone (cf. below).

The carrying power of a voice is not identical with loudness. Often a small or soft voice easily fills an auditorium whereas a large or loud voice is drowned out. A voice carries the better the more fully the vibration is transformed into tone without losing some of its energy to noise elements.

A remarkable standard of loudness is set by the soaring of a well produced human voice above the din of an entire orchestra.

TIMBRE

Remembering that the timbre of a tone depends on the particular constellation of overtones, we readily understand that it is likely to be influenced by almost any particle vibrating in the upper half of the human body. The fundamental vibration of the vocal cords is colored by the secondary vibrations of bones and cavities particularly in chest and head. A good singer can, of course, control these resonances and amplifiers to a certain extent, but an individual's voice will usually be recognized as an identifiable character. The singer's physiology determines, whereas his vocal technique may modify, this character. If a person suffering from a cold speaks with a different voice, one reason lies in the decreased resonance of his sinuses. Conscious control of the sinuses, though possible, is difficult. The easiest modification of timbre is accomplished by the mouth. Sing a certain pitch with a certain loudness on the vowel sound "ee," for instance, and then on the vowel sound "oo." What has changed is the timbre. Each vowel has its own timbre. The vowels, musically understood, determine the timbre of our speech. The consonants contribute the noise element.

A good singer endeavors to be able to keep the same voice character throughout his range without decisive breaks of timbre at crucial points. The energy distribution of the participating partial tones necessarily varies according to the involvement of different resonators and to the physiological condition of the vocal cords at any one moment.

Among the different resonators, the head and the chest constitute two extremes. Accordingly one speaks of a head tone and a chest tone. In a well produced voice, both resonators, adjusting to each other all the time, are heard to function simultaneously. Head resonance favors the high overtones. The vocal cords are permitted to vibrate freely, admitting a wider flow of breath through the opened glottis. The resulting tone shimmers but lacks the ring of the chest register. Chest resonance favors the lower overtones. The larynx is relatively constricted, diminishing the flow of breath through a squeezed glottis. The resulting tone is a bit hard. The correct ratio of these two main resonators to each other accounts for a clear, bright, and free tone.

The vocal cords vibrate similarly to a string. When thick, they favor a stronger fundamental and stronger lower, rather than higher, harmonics. When thin, the fundamental and the lower harmonics weaken whereas the higher harmonics gain in prominence. In the first case, one speaks of a rich timbre, which, however, may sound dull without a sufficient admixture of some upper harmonics. In the second case, one speaks of a brilliant timbre, which, in turn, may sound thin without sufficient support from the fundamental. As the vocal cords thin out with rising pitch and eventually are moreover shortened, the concomitant change of timbre must be controlled by the appropriate adjustment of other participating forces, such as breath and resonators. Most important, however, is the elasticity of the vocal cords, which (like that of any muscle) can be obtained by the proper kind of exercise. The more elastic and supple the vocal cords, the less breath and tension are needed to vibrate the greatest possible mass.

A partial vibration of the vocal cords, from which the fundamental is entirely excluded, produces a falsetto tone, literally translated a 'small false tone.' Among countertenors, one must distinguish between those who really have a naturally high voice and those who sing falsetto.

What musical norms can be deduced from the timbre of a human voice? We may assume that the abundance of vowels in Italian is a desirable attribute of beautiful sound to the Italian ear; the nasal twang of Midwestern American, equally so to the Midwestern ear; the throaty resonance of Russian, to the Russian ear; and the guttural attack of the Chinese, to the Chinese. These norms, which are conditioned by language and probably by climate, are not absolute; but they are reflected by the music instruments that are favored by various societies. Just as our Western culture, in general, is indebted to the Mediterranean tradition, so our musical culture, in many specific ways, has accepted the Italian timbre as desirable. *Bel canto* is not merely a vocal technique but equally so a vocal ideal. Western instruments, on the whole, emulate the *bel canto*. Is there anybody not swayed by the timbre of a violoncello, which seems to "speak like a person"? An Oriental, probably, is not. The violoncello is far removed from his own guttural attack. The instruments he favors, although strange to Mediterranean ears, all sound like emulators of his particular speech. Anyone who has ever heard a Chinese or Japanese orchestra will concede the similarity in timbre between the spoken language and the played tones. Similarly, the nasal resonance characteristic of speech heard in the East is duplicated by such double-reed instruments as the muezzin's *zamr* and the Indian *otu*. Melodies intended to sound Oriental are therefore usually given by Western composers to the oboe or English horn (cf. Rimsky-Korsakov's *The Golden Cockerel*, for instance, or background music to certain movies).

One can only speculate about the reason for a region's favoring one kind of vocal timbre over another. The disguise of a natural voice, in any case, might have a similar function as the wearing of a mask: it is a removal from ordinary reality (one of the basic characteristics of art in general). African feasts have been described where the use of an artificial voice and of an artificial face serves as a protection against the forces of realism.

SOME AESTHETIC CONSIDERATIONS

We live in and by air. The humanly audible is airborne. Singing is a function of breathing. Breathing lies at the root of rhythm and phrasing. An essentially vocal style of music may properly be called "pneumatic," because breath and the vocal organs are here the prime style conditioners.

Against the fundamental activity of singing, the existence of elements properly belonging to an instrumental style should be neither overlooked nor minimized. Wide skips, ornaments, runs—they all suggest features that are quite remote from the pneumatic style. The opening passages of a toccata are a telling example of what is meant. We cannot possibly breathe in empathy with a run; we rather observe an outer happening, a motion, an energetical process manifested in a moving tone. In contradistinction to "pneumatic," the essentially instrumental style could thus be called "kinematic."

Both principles are found side by side in nearly all music, in constant and, more or less, rapid alternation. Vocal coloratura, in the widest sense, is kinematic; and a lyrically "singing" line on any instrument, pneumatic.

The distinction between pneumatic and kinematic styles would be irrelevant, were it not for the difference in our inner attitude toward them. Our musical response instinctively varies with the style; and the consequences range from temporal trends of evolution down to details of interpretation. For the inner acoustics (we should like to use this term in preference to the usual "psychological acoustics" or the now current "psychoacoustics") the distinction is indeed meaningful. In pneumatic music, the circuit voice-to-ear is unbroken. In kinematic music, on the contrary, of which the original domain is instrumental, tone rather becomes an object. Tone is objectified.

At the extreme end of certain contemporary trends, one has come to speak of sounds quite characteristically as "sonorous objects" which are "handled." At the end of the First World War, a frequently encountered term was that of "Spielmusik"—literally, 'music to be played,' but with a frank connotation of 'playful music,' implying the pleasure engendered by the sheer act of playing. This reaction against the high expressiveness of late-Romantic music was carried out under the general catchword of the "New Objectivity" ('Neue Sachlichkeit').

These hints suffice to show the wild sprouting of the kinematic seed in the trail of the triumphal march of instrumental music begun toward the end of the sixteenth century. The cure for an exaggerated application of one principle lies in a call for the opposite one. Our music, diseased by too much kineticism, can be cured only through a return to pneumaticism, that is, to the principles of vocal music.

Exercises

1. Starting on a comfortable pitch, slowly sing an ascending scale (on the syllable "la" or "mi") and notice the physical strain at your extreme top range. Then sing a descending scale and notice, by contrast, the physical relaxation at your extreme bottom range.
2. Sing an ascending fifth legato on the same syllable. Observe the portamento and try to minimize its audibility.

6 | Music Instruments

RELATION TO VOICE

The Latin word *instrumentum* signifies a "tool." The character of musical instruments is thereby well defined. A tool is a mechanical implement to facilitate an operation. It serves as a means to an end. A hammer, for instance, implements effectively the blow of the fist; pliers, the pincer movement of the fingers; a comb, the prongs of the open hand. These tools extend the possibilities of a basic action. Music instruments, similarly, are mechanical implementations of the basic musical activity of producing tones. They are efficient substitutes for the human voice. The task as well as the limitation of any music instrument can be understood first of all by reference to the activity of singing. The voice sets the natural norms which regulate the production and behavior of music instruments. Without these norms, there is no end to the invention of ever new sound-producing contrivances, and no limitation on their behaving as aimlessly as the runaway sorcerer's apprentice.

The supremacy of the voice over instruments is reflected by the development of musical forms in history. Instruments have, of course, existed since the oldest times. The Bible refers in a very early chapter to "the ancestor of all who play the lyre and pipe" (Gen. 4:21). But all evidence indicates that instruments, until much later, primarily doubled in some manner the music that was sung. The service in Solomon's Temple (as we read in I Chron. 25:1-7 and II Chron. 5:12-13) was enhanced by the participation of a chorus, an orchestra, and a band. But all 288 musicians "joined in unison to make a great volume of sound in praising and thanking the Lord." The situation was not much different in Greece. The aulos heard in the theater and the lyra heard in the temple supported the melody sung by the actors or priests. Beside increasing the volume of sound, this doubling must have been a great convenience to singers trying to find their pitch. Even when the instrumental ensemble in the Renaissance began to play music without the participation of singers, most of the musical forms were clearly built on vocal models. The orchestra used its "tools" to perform what had originally been sung. The emancipation of instrumental from vocal music did not really begin in Europe until about 1600. The recency of this date explains the momentum which makes instrumental music appear superior in our present society. There are reasons for assuming that the pendulum has swung as far as it can go. The autonomy of an instrumental style, rightly gained by the masters of the eighteenth century and established by those of the nineteenth, threatens to increase the chaos of the present time unless referred back to its indebtedness to the voice.

The organization of the traditional orchestra still shows the influence of the vocal ranges. The woodwind, brass, and string sections each are composed of four instruments which correspond to the four human voices. Soprano, alto, tenor, and bass are represented, respectively, in the woodwinds, by flute, oboe, clarinet, and bassoon; in the brasses, by trumpet, horn, trombone, and tuba; and in the strings, by violin, viola, violoncello, and bass viol. The various ranges do not exactly coincide, of course, but the relation to the vocal origin is unmistakable.

The case has been argued too often elsewhere to call for further argument in our present context. On the whole, the idea is generally accepted that music instruments are a surrogate of the human voice. This order of values is not contradicted by the recognition that specifically instrumental elements do, indeed, exist. Dance pieces of all times, for instance, seem to be governed by principles that are prevalently instrumental. So are improvisation pieces like toccatas and fantasias, and in general compositions that thrive on rich ornamentation. There exists something like music to satisfy the hands and fingers. The pianist on the keys, the violinist on the fingerboard, the flutist with his pipe, and the drummer with his sticks can leap and bounce in a musical manner that is far removed from the primary vocal experience. Moreover, in the process of emancipation from vocal models, instrumental forms initiated a life cycle of their own like a fresh tree that grows from a cutting. These questions deserve a separate investigation without needing to distract us further at this point. Goethe reconciled the possible discrepancy by suggesting that music instruments, although inferior to the human voice, become equally exalted by sensitive and inspired handling.

INVENTION

The invention of instruments is theoretically unlimited. Any child banging some toys against each other has actually created a new instrument. Practically, some limitations have regulated the invention of music instruments to make them artistically intelligible and usable. Two principles can be isolated.

First, the invention of music instruments has always been a response to a widely felt, genuine musical need. The study of music history acquaints us with the nature of the particular needs, which change with the times. The Greek lyra with seven strings that could not be stopped met the demands of the singer who wanted to double the notes of his confined melody or, as has been suggested, who wanted to draw out the notes of his melody from the total available sonority.[1] The piercing sound of brass instruments—before the days of modern loudspeakers—conveyed signals across considerable outdoor distances; and the hunter calling, the soldier charging, and the magistrate proclaiming were equally satisfied. In the Renaissance, the prevalent musical style demanded a smooth blending of the instruments in an ensemble; as a result, each of the existing instruments developed its own "family"—a group ranging from soprano down to bass and built along the same acoustical principles to ensure the desired mixture. The recorder and the violin appear in complete families to this day; and remnants of this attitude are discernible in other groups, for instance, the oboe, English horn, and bassoon among the double-reeds, or the various tubas among the brasses. The establishment of a public concert life with the concomitant transfer of music from small chambers to huge halls brought about the modification of old instruments and the invention of new in response to the need for enhanced loudness and greater brilliance. This tendency was increased by the individualistic needs of the emerging concert soloist. The recorder was replaced by the transverse flute; the harpsichord, by the pianoforte; the old viol, by the modern violin. The works of

[1] Cf. R. v. Ficker, "Primäre Klangformen," *Jahrbuch der Musikbibliothek Peters für 1929*, pp. 21 ff.

Bach, during whose lifetime most of these changes occurred, are revealing in this respect. Bach experimented repeatedly in his own private way toward meeting the new needs: his inventions of the viola pomposa, the lute-clavicembalo, and a pedal-glockenspiel all point in this direction.

One must not assume that an instrument was invented first to create a need. The problem is old: does the impulse precede the formation of an organ, or will the organ create the need? The precedence of the need over the organ is certainly confirmed by the phenomenon of the human voice, and it is repeated in the historic development of music instruments. In the few instances where the reverse procedure actually occurred, the instrument in question was shortlived: witness the glass harmonica in the eighteenth century, the alto flute in the nineteenth, and the heckelphone in the twentieth.

Electrical instruments provide a recent example of the inverse causal nexus "organ creates urge." They owe their existence mainly to an interest in engineering. This situation is in keeping with nearly all aspects of the present relationship between technology and man, and consequently not peculiar to music. We must distinguish between a class of electrical instruments which purports to imitate traditional instruments, and another class which severs any such connection by proposing new sounds, including noises.

To the first class belong, above all, the electronic organs. They are much less expensive than pipe organs and much more easily installed: in these facts lies their only advantage. They are usable now and then as an ersatz, as in the realization of a basso continuo, as long as one understands that they are entirely different from pipe organs, the sound of which they cannot duplicate any more than even the best phonograph can duplicate an actual performance. The danger here as there lies in the conditioning and concomitant blunting of the musical ear. Persons used to records are often disappointed upon hearing for the first time a live orchestra. Electric organs, like records, may reverse the normal scale of values so that the model eventually is found to be less satisfying than its imitation.

The second class of electrical instruments harbors another kind of particular danger. We mean the catastrophe of the tool becoming the master, and the master becoming the tool. We are told that electronic instruments can produce any sound. In this regard they differ from other instruments, not in principle, but only in degree. Even on a simple string one can produce the most atrocious sounds, as every violin student knows, and quite cheaply at that. The question is: "What do we want?"; and this problem is one of selection—a concept postulated throughout this book. The boundless possibilities promised by electronic instruments offer as annihilating a prospect as does the infinity of possible pitches contained in a string. Limitation is the very condition of art. Where this truth is recognized and a norm imposed upon the technological possibilities, electrical instruments might perhaps be enlisted profitably, and their unusual sounds placed in the service of music. Traditional instruments are close to nature and to psycho-physical norms. Oscillating electric circuits are removed from such an elemental condition and need therefore be "educated" even more strictly than traditional instruments. Furthermore, the naive and grossly exaggerated belief in the inspirational virtue of the process "from outside in" must give way to a renewed belief in the process "from inside out." We should remember that in music, essentially, the urge creates the organ.

The specific need itself for a music instrument can have a purely personal root. One can think of a music instrument as an extension of some inner affinity. A person's feeling for a particular instrument cannot be satisfactorily explained on a mere physical basis. If, as often happens, a musician is highly talented for one instrument and quite ungifted for another, the recognition of the reality of his inner needs becomes more meaningful than any reference to digital dexterity. Parents and teachers of budding instrumentalists should take heed.

The second principle regulating the invention of music instruments derives from the discernment of proportions. The elemental force of proportion, as we have seen, pervades all music. The application to instruments is of a particular validity, for here the idea of proportions and measurements becomes directly a material reality. Proportion determines the whole as well as the details of an instrument—the length, the width, the placement of the holes on a pipe, and even the form of apparently ornamental shapes. Drawings and sketches by the great Cremona violinmakers all testify to their thorough acquaintance with the measure and value of number and proportion. Yet the mere fact that a Stradivari violin cannot be readily imitated should warn us against misinterpreting the idea of proportion as a rigid formula. Instruments builders have gained and checked their norms primarily by ear; they have understood how to apply their knowledge of proportion to the wood and metal in their hands; and their empirical wisdom handed down from generation to generation has proved more fruitful than any mechanical theory. The superiority of the ear over any prescription was already recognized by the Alexandrian scientist Claudius Ptolemy in the second century A.D. when he wrote: "Let us forbear to demonstrate our proposition [concerning proportions] by means of Auloi and Syrinxes, or by means of weights hung on to strings, because demonstrations of this kind cannot attain to the highest degree of accuracy, but rather they occasion misrepresentation upon those who try to do it. For in Auloi and Syrinxes, besides the fact that it is very difficult to discover the correction of their deviations therein, even the points between which the lengths must be measured include a certain undetermined width; and generally the majority of wind instruments have some unordered element added, apart from the blowing in of the breath."[2]

Music instruments are the material realization of an idea. In the process of materialization, the original discernment of proportions may well be enhanced by the variety of shapes that are created. An analogy with biology may be permitted. The principle of life underlies all animals, be they insects or men, fish or birds. One can reach an understanding of all these organisms by proceeding from a basic natural principle. One will, inversely, increase one's insight into this principle by studying the variety of material forms. The monochord suffices for a study of basic musical principles. Wind instruments, by their very nature, will enlarge the scope of study. Overtones, for instance, though demonstrable on a monochord, can be "discovered" and investigated more naturally on an air column than on a string. The longitudinal vibration of pipes contrasts with the transverse vibration of strings. The sound of reeds has its own subsidiary, but particular, laws. The invention of instruments thus increases our insight into the force and application of proportion.

[2] Here quoted from *The Greek Aulos* by Kathleen Schlesinger (London, 1939), pp. 130 f.

CLASSIFICATION

The existence of a multitude of music instruments gives rise to the demand for a suitable classification. We know that classification, by definition, is one-sided; for any one classification will fulfill certain functions but not others. Today, as in the time of Beethoven, common practice refers to woodwinds, brasses, percussion, and strings. This classification, though it has served its purpose, can be attacked on several grounds. The terms "woodwinds" and "brasses" refer to the material of which the respective instruments are built; "percussion," to a manner of playing; and "strings," to the source of the vibration. Apart from the fact that this mixing of categories is illogical, the terms themselves are by now incorrectly applied. The flute and saxophone, for instance, are grouped with the woodwinds although flutes are nowadays made of metal, and saxophones always have been. Both the organ and the piano are keyboard instruments, but the one could be classified with the winds, the other with the strings; and the piano, moreover, could be called a percussion instrument. To satisfy varying practical and intellectual needs and purposes, several classification systems of instruments are not only possible but necessary.

In the course of time, different musical principles have been used to classify instruments, the stylistic concern of a period justifying any apparent prejudice. A tendency of the late Middle Ages is documented according to which music instruments were grouped "tant hauts que bas"; and sonority—loudness as well as timbre—probably formed the basis of this aesthetic distinction. Renaissance musicians preferred an arrangement by pitch. Flute, violin, and trumpet were assumed to belong together because each could play the soprano part. For analogous reasons, bassoon, violoncello, and trombone characterized another class. This system—strange to us—made a great deal of sense at a time when instruments were primarily employed to double, or at least to emulate, vocal lines. What mattered to a musician was to know which instruments could be utilized for a particular voice range; and compared to this pragmatic concern, it was irrelevant whether the suitable instruments were of wood or brass. Actually, arrangement by pitch is still the governing principle for writing orchestra scores, from the piccolo down to the double bass, although other factors modify it.

When the practice of thorough bass dominated the style, the primary question was the suitability of a particular instrument for the melodic lines or for the harmonic accompaniment. Proceeding from this stylistic consideration, Agostino Agazzari, in an essay of 1607 entitled "Del suonare sopra il basso con tutti strumenti e uso loro nel conserto" ('How to play above a bass with all instruments and how to use them in concert'), accordingly grouped instruments into "fundamental" and "ornamental." Fundamental instruments were those that could guard and hold together the whole body of sound, in short, harmonizing instruments like the organ, cembalo, chitarrone, lute, and others. Ornamental instruments were those that made the music more pleasant and agreeable by playful embellishments of the contrapuntal lines, in short, melody instruments like the violin, cornet, and also the lute when used as an ornamental instrument. This kind of distinction will be readily understood by every jazz player, who accepts his classification as the player of a harmonizing instrument (often the guitar, or the piano, or the lower wind groups) or of a melody instrument (e.g., clarinet, trumpet), besides recognizing rhythm instruments (be it the drums or the double bass) as a third category.

When Bach chose the collective title "Clavier-Übung" for a variety of compositions to be played on the clavichord, harpsichord, and organ, he applied, not so much a purely musical, as a technical principle of classification. What links these three instruments (to which the modern piano must be added) is not any acoustical quality they might share but a mechanism. This kind of order was irrelevant to Berlioz in the nineteenth century when timbre became an important aesthetic factor.

A modern classification proceeds from the vibrating, sound-emitting body. First submitted around 1888 by the custodian of the thousands of music instruments at the Brussels conservatory, Victor Mahillon, it was developed around the beginning of the First World War by Erich von Hornbostel and Curt Sachs in Berlin. This system, which is neither better nor worse than any other, distinguishes among four classes of music instruments: chordophones, aerophones, membranophones, and idiophones. The meanings of the names are evident from the Greek roots. In a chordophone, a string (Greek *chordé*) is the source of the sound (Greek *phoné*). In an aerophone, it is a column of air (Greek *aer*). In a membranophone, a stretched skin (Greek *membrana*) sets off the primary vibration. In an idiophone, the elasticity of the instrument is capable of rendering sound by itself (Greek *idios*). The addition to these four classes of a fifth has recently been suggested: electrophones, in which an electric vibration results in sound. Electrical instruments, however, could be grouped with the membranophones because they depend on the membrane of the speaker to transform the electric current into sound.

To the musician this classification is less gratifying than to the physicist because of the inherent materialism of the approach which does not admit of any value distinctions. Here, as in all other possible classifications, one finds strange bedfellows. Recorder and tuba, violin and piano, harp and violoncello, triangle and xylophone—each of these pairs holds together by an intellectual rather than a musical affinity. Among the wind instruments of the orchestra, moreover, only the flute is an aerophone in the precise sense; for in all others, the reeds or lips, as the case may be, are the source of the vibration.

MATERIAL AND FORM

The relation of sound to material is central to an understanding of music instruments. The main conditions for the creation and enhancement of harmonic vibrations are elasticity and inertia. Without them, a string or reed cannot vibrate adequately and commensurately to the production of a musical tone, and the soundbox cannot respond properly and sympathetically to the initial impetus. The material used in the making of music instruments may be judged (1) in regard to its practicability, (2) in its role as a participant in the production of timbre, and (3) by its intrinsic effect upon the human being.

(1) Every material has a style inherent in its properties. It demands to be treated according to its own nature. It consequently favors the creation of some forms over others. An animal skin, for instance, is readily cut and stretched; and the roundness of a drumhead answers the need for evenly distributed tension. The malleability of brass makes it best suited to imitate the prototype of wind instruments offered by curved animal horns. It further permits the bending and coiling of tubes too long to handle, as best illustrated by the otherwise unmanageable length of about twelve feet of the F-horn.

A given material may, through art, be forced into shapes properly belonging to another material. Wood can be bent, to a degree, with the help of steam and pressure. The serpent, a bass instrument popular in the sixteenth century, was sometimes made of wood. Special precautions, however, had to be taken at the critical points: sinew bindings were applied, and the whole instrument clad in leather. The alpenhorn (which in its present form looks like a huge smoking pipe rather than an animal horn) consists of ten-feet-long wooden staves bound together by strips of birch bark. From a practical viewpoint, it could certainly just as well be made of metal.

Some construction principles may be embodied in one material as well as in another. A piano frame, for instance, lends itself equally well to wood and metal. The latter can be made much stronger for the same dimensional specifications, but the basic form remains the same.

The natural use of the material, subject to the minor qualifications just mentioned, yields a practical form when adapted to the needs of the player. A string must not be of disproportionate length; and the minimum soundboard will have the shape of a long narrow rectangle. The god Pan tied together pipes of different length for his syrinx; the drilling of holes in *one* pipe effects the same results with greater economy and versatility.

(2) The questions of the influence of material on timbre has been, and still is, the subject of endless discussion and experimentation, spurred mainly by the changeover from wood to metal flutes, and from gut to steel strings. As of today, the results of scientific investigation are not conclusive. Research has relied either on an analysing apparatus or on subjective reactions. Experiments with sound analysers rest on the postulate that an apparatus is the equivalent of the ear, whose functioning and interpretative potentialities are not even entirely known. Polling experiments depend on the auditory and musical qualifications of the subjects, and the results lack the convincing force of certitude. Thus both methods are open to skepticism.

Investigating the influence of material on timbre, one might do well to bear in mind the different tasks expected of the material. It may be the source of the vibration, or enhance the primary vibration as a resonator, or serve primarily as a physical boundary. In idiophones, true to their name, the material itself is agitated. Triangles, gongs, and bells are metal instruments by definition. The specific timbre of the xylophone demands wood. The requirements of sounding plates, as in the celesta, can be met by both glass and metal. Timbre differences between gut and steel strings are more apparent when the string is plucked or struck rather than bowed. Gut strings have therefore remained essential for the timbre of the harp but have yielded readily to steel strings on the violin. One can easily presume a gut-stringed piano to exhibit a timbre very different from one obtained by struck metal strings. Soundboards and soundboxes are traditionally of wood, and no other material seems fit to resonate with equal success. The organic properties of wood apparently have a special bearing on timbre which cannot be imitated by inorganic matter. Common terminology recognizes the inherent truth of this basic distinction by defining as a positive value the "life," the "liveliness," of a tone. The dividing line between the function of the material as a resonator and as a physical boundary is not always easy to draw. Do the walls that contain the air column in wind instruments participate in the formation of timbre as resonators or as mere enclosures?

Much can be learned about the influence of material on timbre from the experience of organ builders who have attempted to influence tone by carefully experimenting over the centuries with the weight, density, rigidity, and potential resonance of pipe walls. At one extreme, there are the heavy lead pipes of the Italian late Renaissance. The thick, inert material encloses the air column like a vessel, without significantly participating in the vibration. The resulting tone carries well and is strong relative to the initiating force. At the other extreme, there are the thin-walled wood pipes of an *organo di legno*. The highly resonant wooden walls of, for example, Lebanon cedar (not to be confused with the American red cedar) vibrate along with the air column. This process costs energy, for the air column loses vital force to the yielding, excitable wood. The tone is weakened, but it assumes "wood qualities," that is, the partial tones acceptable to the pipe wall (normally the fundamental and a few low overtones) are radiated from it as from a soundboard. Compared to lead pipes, wooden pipes mellow the tone in a manner that is suitable for chamber music but that would be incongruous in a big hall.

Between the two extremes, tin pipes have proven their adjustability to varying acoustical needs. The quality of thick tin approaches that of lead. Thin tin pipes are made of hammered plates, the internal tension of which reinforces certain high partials. Gottfried Silbermann, who built fine organs during Bach's lifetime, characteristically liked to mix on the same instrument thick-walled, widely proportioned pipes on the great and pedal with thinner, pene-trant pipes of hammered tin on the swell and choir.

(3) The intrinsic effect of material upon man is less easily exposed than practicability and timbre formation. Compared to considerations which are mechanical and practical, on one hand, and musical and organic, on the other, the metaphysics of materials, as it were, has no obvious, undisputable bearing on our appreciation of instruments and their sound. Yet ignoring it would be a mistake. The nature of stone, or metal, or organic material, we submit, evokes a certain reaction in us which is particular in each case. People of former times were more sensitive to these material distinctions than we are today. Attitudes toward metal illustrate the point. In all civilizations it was considered the product of subterranean forces, rough and ugly like Hephaestus and Albe-rich. Metal means destruction and warfare. Ovid speaks of metal as the root of evil, and of "the guilt of iron, and gold, more guilty still." This appraisal is deeply engrained in us. Just imagine a living room or a bedchamber lined with metal. The very thought of it makes us shudder; and this aversion is not satis-factorily explained by any such factor as the coldness we might feel when touching metal. Indeed, it is quite another coldness from which we here shrink, and one defying ordinary references to thermal conduction.

By a reversal of symbolism—a familiar phenomenon—metal can also spell glory and magnificence. In this sense only, brass instruments are sometimes admitted into religious services; but they are never associated with religious devotion.

The replacement of the original animal horns, which were aggressive weap-ons, by brass seems appropriate; and we cannot quite forget the hollow reed as the prototype of woodwind instruments. But there is something ridiculously grotesque about the idea of a papier-mâché or plastic trumpet, even if the sound could exactly duplicate that of the metal instrument. We can easily imagine that one day all aerophones might be metallic and all membrano-phones plastic—a culmination of the general attitude of our age; but the sheer presence in the composer's mind of all-metal winds and all-synthetic drums would surely change the essence of the music.

The predominance of brass instruments in the modern orchestra—doubtless related to the current times of aggression, war, and turmoil—parallels that of steel in modern buildings. In the construction of Solomon's Temple, the use of metal tools was expressly forbidden (I Kings 6:7). Chartres rose without metal. The general promotion of metals has much to do with the acknowledged deterioration of architecture as an art. Nobody sensitive to quality, to the life of an artwork, can prefer a metal building to a wood or stone house.

Thus far in our discussion, the concept of form has appeared only to the extent to which it is inseparable from the concept of material. Beyond this connection, however, form or shape has a life of its own which, properly understood, is to a degree independent of the idea of matter.

There seem to be two kinds of factors, apart from those connected with the material, that have shaped music instruments. One kind derives from the intention to favor the immanent possibilities of tone. To this class belong most obviously the shapes of soundboxes, foremost those of the violin and lute families, and of bells of wind instruments. The other kind has a geometric origin, be it the product of playful fancy or the expression of symbolic significance. To this class belong, for instance, the shape of the "F" hole on the violin and the precise circular coiling of the sousaphone.

The form-giving factors of these two classes have sometimes been opposed as "artificial" to the "natural" form principles of animal horns and reeds. This distinction makes good sense when one proceeds historically. Animal horns and bones as well as hollow reeds, without doubt, provided the first wind instruments and have remained formal models ever since. The situation is less clear when we turn to stringed instruments, for natural strings offer themselves less readily. There is real danger, however, in contrasting "natural" and "artificial" principles in the realm of shapes, because the underlying idea of geometry is the same in both cases. In relation to the enormous wealth of natural forms, artificial forms created by man are not an opposition but a continuation. God created the world so that man may carry on creation. The contrast loses further meaning when we reflect that man chose, not any natural object or even any horn, but only such a one as serves his purpose well.

In woodwinds, the shape of the air column as defined by the bore remains hidden. Hence there is little in the outward appearance that seems to be fashioned merely to favor the immanent acoustical possibilities. The outward shape of woodwind instruments is characteristically simple. The important principle is the containment of an air column by a tube of a certain internal shape. When the tube reaches too great a length for handling, it is bundled, as in the bassoon. The bore in brass instruments, on the other hand, fashions the outside appearance, or rather, is fashioned by the visible side of the wall. The outward shape of brass instruments, aided by the malleability of the metal, thus acquires an expressiveness of its own which is often inspired by geometry. Neither the circular coiling of the horn nor the oblong one of the tuba is dictated by any practical necessity. The tremendous bell flare in some brass band instruments serves no other purpose than to impress the beholder with a suggestion of power.

The most elaborate, most exquisite forms are proffered by stringed instruments. The form of the instruments of the modern violin family (violin, viola, violoncello) appears suddenly around the middle of the sixteenth century, associated with such master craftsmen as Caspar Tieffenbrucker and Gasparo da Salò. Form is not the only sound-shaping factor—wood, varnish, and craftsmanship immediately come to mind—but it is doubtless of basic importance. The form of the violin, crystallized out of a great many experiments with different shapes, is particularly puzzling because of the amazing efficiency of the soundboard, on one hand, and the apparent impossibility to explain it rationally, on the other. Hans Kayser submits an interesting thesis in a monograph devoted to the problem.[3] He develops the shape of the violin body with the help of a peculiar curve obtained from the transformation of the tone ratios into vectors. The results are remarkable in more than one respect. Kayser's curve explains, for instance, why the end blocks and corner blocks do not interfere with the body resonance. Furthermore, the usual scroll shape of the head appears, not as the product of whimsical fancy, but as an embodiment of the very curve that determines the body form as a whole. Perhaps most interesting of all is the observation that Kayser's construction principle not only allows for the stated variations in the practice of the master builders but actually opens the door to hitherto unexplored variations and even new forms.

What, one might ask, is the value of geometry and symbolism in music in-

---

[3] *Die Form der Geige: Aus dem Gesetz der Töne gedeutet* (Zürich, 1947).

struments when it cannot be heard? Only modern man could ask such a question. In times when man was "whole" (in the sense of both healthy and unbroken), "functionalism" (if the concept had then existed) could only have been understood as a function of the whole man, himself a function of the cosmos. In this spirit, the external appearance of a music instrument had to match the significance of music. It had to be "in tune" with its entire function, which went far beyond simple tone production. If this sounds strange, remember the singular satisfaction we still obtain from handling and reading a book in which the beauty of printing and binding matches that of its content. This connection between content and form holds true of instruments in an even more intimate way.

MUSICAL CHARACTER

We know that the wave motion produced by a vibrating string is transverse, and that the wave motion of air in a wind instrument is longitudinal. The qualitative, musical distinction between stringed and wind instruments, however, has roots that lie in another sphere. Let anybody examine his own reaction while listening to a band, and then while listening to a string orchestra. The difference between the two experiences, even if both groups were to play the same composition, finds no explanation in the nature of the wave motion. The Greeks were more conscious of this distinction than modern man seems to be. We can do no better than follow their thoughts.

The lyra and the aulos were the two most important music instruments of the Greeks. The lyra had strings which were plucked or struck. The aulos was blown through a double reed. The lyra was the instrument of Apollo, dedicated to quiet worship in the temple. The aulos was the instrument of Dionysus, employed in the ecstasy of the dithyrambus. The lyra and the aulos embodied the polarity of two elementary apperceptions of music, which between each other held the possibilities of all musical experiences.

The invention of the lyra is ascribed by mythology to the god Hermes, who gave it as a peace offering to Apollo ("Homeric Hymn to Hermes"). Hermes had the idea when he found a tortoise and deliberated that the shell could produce sound:

"And though it has been said
That you alive defend from magic power,
I know you will sing sweetly when you're dead."

He kills the animal and "at proper distances" stretches "symphonious cords of sheep-gut" across the hollow shell. The point of the story is the identification of the invention of the lyra with the discovery that the world has a sound. The leading authority on Greek music, Thrasybulos Georgiades, bases the effect, the magic charm, of lyra music on the awe at the realization that an object can give forth sounds. Moreover, the several strings of the lyra produce tones that fit to each other. The phenomenon of consonance emerges from the music of the lyra and makes man conscious of the "sound of the world" around him. Music on the lyra is not a personal expression but a reflection of the harmony of the world.[4]

The invention of the aulos by Pallas Athena is described by Pindar (in the Twelfth Pythian Ode). After Perseus had decapitated the Medusa, Athena heard the wailing of the Gorgon sisters.

She invented the full-sounding melos of the aulos
To imitate with this instrument the wailing clamor
That grew from the mouthing jaws of Euryala.
The Goddess invented it and gave it to mortals to use.

The art of playing the aulos is described in this story as the musical rendition of a human cry. In the course of the Ode, Pindar distinguishes between suffering and a spiritual interpretation of this suffering. The one leads to a personal outcry expressive of human life. The other lends an artistically objective form to suffering and becomes a divine, liberating, spiritual act. Man turns spiritual when he receives the divine gift of the aulos from the hands of the goddess. Pindar is very concrete in identifying the music of the aulos with an expression of human emotion. The tone of a wind instrument is similar to that of the human voice, for both are created by the pulsation of breath. The word "aulos," it is worth noting, is derived from the onomatopoetic verb *auo*, 'to shout.' The

[4] Thrasybulos Georgiades, *Greek Music, Verse, and Dance* (New York, 1956).

aulos and the human cry both express the feelings of the individual. In character, this kind of music is necessarily monodic. It does not relate to the outside world but rather centers on the personal passion. It is linear.

Lyra and aulos, strings and winds, thus stand for two extreme archetypes of musical instruments. The lyra connects music with the objective laws of the cosmos. The aulos conveys a subjective expression of human life.

Against the musical superiority of both stringed and wind instruments, the prevalence of noise sets apart percussion instruments of both definite and indefinite pitch. Noise assumes an essential role in musical structures only when serving a rhythmic function. A theory of rhythm transcends a theory of musical acoustics, and we shall therefore not speculate on the rhythmic relevance of percussion instruments. But a word is in order on noise as a purely acoustical ingredient of musical relevance.

We can isolate the noise factor from rhythm by thinking of a continuous noise. Timpani rolls are a common feature of orchestra music. Their effect is always meaningful in a particular way. They are threatening. One begs the question by relating the experience to the distant rumble of thunder. Why is thunder generally frightening, although one usually has nothing to fear while hearing it? Both the thunder and the timpani roll, compared to a noise-free musical tone, are symbols of the undistilled outer world. The experience is analogous to the one brought about by the glissando (cf. pp. 5 f.). The uncontrolled dynamism is an outburst of the brutish world which, though having produced us, remains our antagonist. The life-threatening quality of noise disappears in the ordered realm of tone. In music we feel "at home," because tone is a refinement, a distillate. We experience comparable pleasure looking at a crystal, another distillate from nature.

Why is a triangle tremolo, which is also continuous noise, less frightening than a drum roll? First of all let us assert that the sound of a high-pitched whistle or of a persistent doorbell is startling. The least one can say about a triangle tremolo is that it adds excitement to the orchestral texture. The main reason for the difference, however, is the suggestion of increasing mass with falling pitch. The peculiar aesthetic meaning of noise reveals itself more characteristically in deep sounds. The low-pitched noise of the timpani and of thunder is fearful because the immense mass threatens—more effectively than the lighter triangle—to crush the beautifully ordered, rare distillate which we treasure.

The low pitch of musical tones partakes of this suggestion. Verdi uses an extensive bass-viol solo to accompany Otello's murderous entrance into Desdemona's bedchamber. The effect would be lost if the same passage were played by a flute. By moving the mass of deep sound out of the irrational region of noise into the rational reign of tone, the bass viols, while expressing a threat, accomplish an artistic purification.

Exercises

1. Select a group of instruments (such as children's instruments, medieval instruments, Oriental instruments, etc.) and classify the instruments within each group according to different applicable principles.

2. Set up pairs of instruments which belong to the same family according to one kind of classification, and to different families according to another kind of classification.

3. Take cardboard (mailing) tubes and metal (steel or copper) tubes of the same length but of various diameters. Blow across one end and listen to the difference in timbre between tubes that are narrow or wide, cardboard or metal, open or closed at the end.

4. Play a recording of an instrument that is available to you (e.g., piano or violin). Try to regulate the timbre and loudness of your phonograph system to sound as nearly as possible like the live model. If you have a tape recorder, make recordings of single tones, scales, and simple melodies; experiment with the position of the microphones and the settings of the controls so as to imitate the natural sound as nearly as possible. Observe the remaining differences.

7 | Stringed Instruments Other than Keyboard Instruments

CONSTRUCTION

Because our present concern focuses on tone production by the player, the individual variables among the many members of the string family may be slighted in favor of the constant characteristics that are shared by all. The important feature is the vibration of one or more strings, which is caused by plucking, bowing, or striking.

A vibrating string displaces so little air around it that a soundboard is necessary to enlarge the motion and consequent disturbance. To this end, stringed instruments employ a hollow wooden box. Beyond being an amplifier, this soundbox determines to a maximum the quality of the entire instrument. The resonance of the soundbox—the product of material, shape, and proportion—creates the desired musical value. Peak resonances must be avoided, for the favoring of one pitch over all others could only be disturbing; the phenomenon, if occurring, is appropriately referred to as a "wolf." The range of possible resonances makes a stringed instrument sound rich or dull, sensitive or stiff, expressive throughout the total range or unevenly responsive. A soundboard has one or more holes cut into it—in the shape of the letter "F" on the violin, and round, often covered by a rose, on the lute. This hole breaks the rigidity of the material so that the board may vibrate more easily. It also permits the air inside the box to communicate the vibration more readily to the air outside the box.

The vibrating strings may be stopped by the fingers of the player, as on the violin, or by a more limited mechanical device, as on the harp. They remain entirely open, vibrating always at their full length, on other instruments such as the psaltery. A third group combines both types so that the player may finger melodic progressions on some of the higher strings while using lower open strings for bass support (zither, Renaissance lute). These varying techniques have an obvious bearing on the number of strings employed, for a stopped string can perform the duties of a multitude of open strings. One notes that the number of strings is 4 on the violin, viola, and violoncello; 4 or 5 on the bass viol; 6 on the guitar; 11 on the Renaissance lute; and 46 to 48 on the harp.

The violin family utilizes a bow to initiate the vibration without, however, sacrificing the method of plucking common to most other stringed instruments. The eminent musical role of the violin family may well be explained by the maximum control available to the player over pitch, loudness, timbre, and duration—in short, over the total tone production. No other instruments offer the performer a comparable accomplishment.

PITCH

On the monochord, we learned that a shortening of the string raises the pitch. Physical experiments readily show that other related factors enter. The length remaining constant, a heavier string sounds lower than a lighter; and increased tension, which can be measured by a weight attached to a string, raises the pitch. The French philosopher and scientist Marin Mersenne, a Franciscan monk, formulated these laws, which have been appropriately named after him, in his magnum opus *Harmonie universelle* of 1636-37. They may be summarized as follows:

The frequency of a vibrating string is:
1. inversely proportional to its length;
2. directly proportional to the square root of its tension; and
3. inversely proportional to the square root of its weight per unit length.

In practical terms, a string player gains a higher pitch by (1) shortening the string, or (2) tightening the string, or (3) using a thinner and lighter string. In mathematical terms, the higher octave 2/1 sounds if (1) the string is halved, or (2) the tension is quadrupled, or (3) the weight is reduced to one-fourth.

The string player is concerned with the weight of the string only when he buys it and mounts it on his instrument. The lower strings are heavier. On the harp, they are both heavier and longer. Strings of identical weight and length can be tuned only to pitches that lie near each other, for the tension increasing by the square would soon snap them altogether.

The string player tuning his instrument applies the Second Law of Mersenne. This fact is obvious on the violin, for instance, where all strings are of the same length. He changes the tension of each string by turning a peg attached to one end of it, always carefully remembering that a considerable turn is necessary to accomplish a minimal adjustment of pitch. The particular tuning of an instrument is founded on both organic norms and man-made conventions. Tuning the strings on the violin, viola, and violoncello in fifths is convenient for the player's ears because he can check the correctness of his effort by referring to the relatively loud 3rd partial tone of the next lower string. Furthermore, it is convenient for the player's hands, because he has just enough fingers to play the scale tones that lie between two adjacent open strings. On the bass viol, the overall length of the instrument also proportionately increases the distances between distinct tones on any one string. The move from 3/4 to 2/3 of the total string, for instance, that is, a whole tone, covers about 3½ inches on a bass viol as compared to about 1 inch on a violin. The size of the human hand remaining the same, the strings of a bass viol are accordingly tuned in fourths; overtones may again be employed to check the purity of the intervals. Guitars are often more arbitrary in their tuning, although the strings usually follow each other in fourths with a major third somewhere in the middle. This arrangement permits the player to sound full triads without having to shift much the position of his left hand. The harpist tunes his instrument in a diatonic major scale across the almost seven octaves at this disposal.

On the violin (we shall henceforth use this term to include viola, violoncello, and bass viol, unless otherwise qualified) and guitar, the player's ears and hands perform all the operations familiar to us from the monochord. To change the pitch on any string, he divides the string by stopping it with his finger at the appropriate spot. At 8/9 of the total length he produces the whole tone above the open string, at 4/5 the major third, and so forth. This division is precisely performed, not by a measuring tape, but by an evaluating ear. The musician checks the correctness of the division by hearing it.

He knows that he has halved the string when he sounds a pure octave. Actually, violinists do not need to think of the underlying division. To them, the value of the tone has a primary reality in the totality of the tone-number. Measure, which tyrannizes the modern world, assumes a complementary role in the musical experience.

The division, performed entirely by ear on the violin, is made easy on the guitar by ridges on the fingerboard which mark the resultant spots. These ridges, called "frets," occur also on other instruments, for example, the viola da gamba. The musical loss outweighs the mechanical gain; for the performer is no longer free to make the minute adjustments in intonation that distinguish his sensitivity. Frets are really an elevated projection of the markers of a measuring tape on the fingerboard.

A pitch question is caused by the vibrato that is practiced by most performers on stringed instruments. This familiar effect of an expressive quiver is in reality a pitch pulsation. If the vibrato becomes either too slow or too wide, the substitution of numerous tones for one characteristic pitch loses virtue as it gains audibility. In this regard, a vibrato, whatever its aesthetic advantages, permits a player to obscure the exact pitch by oscillating among a variety of pitches in the neighboring area.

Does the pitch of a string remain the same whether plucked or bowed? One might assume that the vibrations of the bow added to those of the string would influence frequency. This, fortunately, is not the case. The bow is so constructed that it does not set off a definite periodicity of its own, and the bowed string remains unhampered to pursue its free vibrations.

The harpist alters the pitches of his given diatonic scale by a singular pedal mechanism. All strings of the same letter-name (e.g., all B-flats) are hooked to one pedal, so that the player deals with a total of seven pedals. When a pedal is depressed, it tightens all the strings connected to it and thereby raises the pitch. A pedal can be locked in two different positions, of which each raises the pitch by one half-tone. Because the pedals can only raise, but not lower, the pitch of a string, the basic diatonic scale on the harp is tuned in C-flat major. By depressing the C-pedal once, the harpist alters all C-flats on his instrument to C-natural; by depressing it one more step, to C-sharp. Thus all keys and tones can be accommodated; but only those glissandi are possible that dispose of all seven strings, however tuned, within every octave. Here are some specific harmonic formations that are realizable, among others, as a harp glissando:

Figure 34

LOUDNESS

While the violinist's or guitarist's left hand finds the desired pitch, his right hand regulates the desired loudness. In pizzicato, the situation is identical with the one experienced on the monochord and described earlier. The amplitude of the vibrating string is the only relevant factor. The farther the finger removes the string from its equilibrium, the louder the particular tone will sound; and neither tension or speed of the finger has a bearing on the resultant dynamics.

The situation is less simple when the right hand, instead of plucking, guides a bow. The bow grips the string and drags it along on its path. Rosin rubbed on the bow hair increases the friction and strengthens the grip. At a certain point, when the resilience of the disturbed string becomes greater than the force of the dragging bow, the string slips back, overshoots its normal position, and is then again quickly seized by the bow so that the process begins anew. The amplitude of the vibration is thus related to the force that impels the string away from its normal position.

The violinist generates this force by the pressure and the speed at which he pulls the bow across the string. Judicious management of these two factors is necessary for satisfactory control over dynamics. The participation of pressure and speed must be well proportioned, for evidently neither pressure alone nor speed alone produces a tone. Beginners usually rely on too much pressure and therefore, in the fiddler's jargon, "scratch." Advanced players who indulge in too fleeting a bow lack dynamic intensity. The exact proportion of pressure and speed is a primary task for a string player to learn but a musically futile one for an acoustician to calculate. The ear, here as elsewhere, is the quickest and most precise judge. It guides the arm of the player who adjusts bow pressure and bow speed against each other and toward the particular demands of every note. Nothing else influences the amplitude of the bowed string.

A musically precarious situation arises when two violinists play in unison. Each player's bow grips and releases his string in a continuous alternation of phases. The possibility that the phases of both bows will exactly coincide is minimal. As a result, the two tones reinforce each other at one instant and interfere with each other at the next; and even the best two violinists playing in perfect unison thus produce a sound of undulating loudness. This tremolo is made less obtrusive by the use of at least four players for the formation of a blended and balanced unison section.

TIMBRE

As the string is set in vibration, two variables controlled by the player significantly affect the overtone constellation. One is the place at which the string is touched; the other, the precise manner in which finger, plectrum, or bow are applied.

The principle underlying the first variable is simple enough. The vibrating string forms loops and nodes corresponding to the various harmonics. A string excited at a loop favors the particular harmonic. A string excited at a node eliminates the particular harmonic.

We know that the midpoint, for example, of a string determines the octave. There the second partial has its node. If the player excites the string at this point, he practically eliminates from the overtone series the octave with all its subsidiaries, in short, all even-numbered partials. Because upper octaves brighten the fundamental, a violinist exciting the string at its midpoint produces a somewhat hollow sound (which may be musically appropriate at times). Similarly, the player touching the string at one-third of its length suppresses the fifth without interfering with the octaves. The resultant timbre has its own characteristic.

The player does not consciously divide the string to produce a particular timbre. His ear soon tells him what place on the string corresponds to the tone color he desires. In general, he bows the string at about one-tenth of its total length. Upper partials gain ascendancy over the fundamental as he moves nearer the bridge. The indications *sul ponticello*, which produces a thin glassy timbre, and the other extreme "on the finger board," which sounds colorless, are often prescribed by composers.

The second variable—the manner in which the string is agitated—is empirically known to all string players. The significant distinction between a plucked and a bowed tone can be related to the quick vanishing of the colorful high harmonics after the initial impact of the pizzicato, as compared to the continuous strengthening of the same harmonics by the movement of the bow. In this respect, loudness has a bearing on timbre; for the relative strengths of the partials change with the amplitude. Whereas the player immediately loses his control over a pizzicato tone, he cannot help varying, if ever so minutely, the strength of a sustained bowed tone (nothing is more difficult —practically barely possible—than an absolutely even violin tone). He thus necessarily affects the timbre by every dynamic fluctuation of the fundamental with all its tributary overtones.

Both the finger and the bow, moreover, have a considerable width, which may interfere more or less with the form of the vibration. The wider the part of the string covered by the plucking finger or the bow hair, the more the higher overtones corresponding to that part are dampened; for the width of the finger or the bow, we may imagine, smoothens the curve formed by the vibrating string. A violinist accordingly changes the amount of bow hair that touches the string in order to influence the timbre. Playing with the thin edge of the bow hair seems an extreme device for favoring the high overtones; but it can be further surpassed by playing with the wood of the bow, *col legno*, an occasional practice in twentieth-century compositions.

The violinist's left hand, too, can have a bearing on timbre. If he touches the vibrating string lightly at a point instead of stopping it, the partial for which that point is a node will sound in complete isolation. The shape of the vibration approaches a pure sine curve. The resulting timbre is nearly bare of overtones, a bit eery, and often prescribed by composers. Such tones are called "flageolets" or simply "harmonics." A string thus touched vibrates as a whole but in aliquot parts. As a result, the pitch of a flageolet tone corresponds to the shortest of these uniform sections regardless of which of the nodes is touched. Divided at 1/2, the string produces the octave as a harmonic; at 1/3 or 2/3, the twelfth; at 1/4 or 3/4, the second octave; at 1/5, 2/5, 3/5, or 4/5, the major third two octaves up; etc. To predict the pitch of a harmonic, one need merely identify it with the denominator of the string fraction while assuming the numerator to be always 1. Open and stopped strings alike lend themselves to the production of harmonics, which are generally distinguished by the names, respectively, "natural" and "artificial," although both are quite natural.

The harmonics actually used on stringed instruments are a selection from a great number of possibilities, which can be readily demonstrated on the monochord. (In experimenting with the monochord, dampen all strings except the one involved; otherwise the harmonic from the one string is likely to be covered up by the resonance of the other strings.)

Particularly interesting is a series of harmonics which might appear to consist of undertones of some very high overtone but actually must be recognized as a deviation from a strict overtone series. This kind of deviation involves the concept of "tolerance" which is inseparable from the phenomenalized world (cf. pp. 185 ff.). On the string of our monochord (c_1 = 120 cm), the following (and subsequent) harmonics can be produced apparently as undertones of 1/64 (from the fifth undertone on, the first four being too high) but actually as overtones of 1/1:

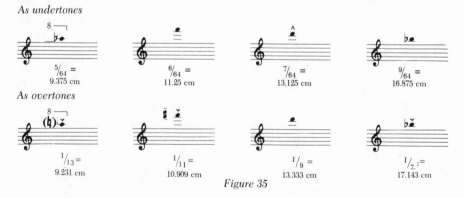

Figure 35

The body of the violin not only reinforces the vibrations of the strings by means of forced resonance but also enriches them by adding its own free resonance. These sympathetic vibrations of the violin body usually lie between 3000 and 6000 frequencies (approximately in the fourth and fifth octaves above middle *c*) and in their totality are known as the "formant." The formant of a viola lies lower than that of a violin, because of the larger body of the viola. The timbre difference between the same note played by the one and by the other instrument results from the reinforcement of a group of higher harmonics by the violin formant than by the viola formant. The suggestion has been made that the formant is the main criterion that distinguishes a Stradivari from a common violin. This suggestion, however, begs the question *how* Stradivari achieved a more sympathetic violin body than other builders.

Be it as it may, the body of a stringed instrument influences the timbre decisively. Amidst the many theories (based on varnish and other factors), the significance of proportion and material emerges from the sketches of the Cremona masters as much as from our own deductions. The fact that copies of great violins are not successful merely proves the point that none of these theories can ever be reduced to a mechanical formula. Stradivari himself must have known and experienced the distance between the idea of an instrument and the particular incarnation: the asymmetry of most of his instruments expresses his respect for an indomitable life force, be it only that of a knothole.

Exercises

1. Experiment with various sizes and shapes of soundboards and soundboxes by fitting a string (or strings) to boards of different sizes and shapes, tables, cigar boxes, metal boxes, plastic boxes, etc.
2. Observe timbre differences between gut and steel strings—plucked, bowed, and struck. Maintain the same point of attack for your comparisons. Then observe the effects after changing your point of attack.
3. In the following exercises with flageolet tones, check on the monochord all the results first obtained by arithmetic:

    a.   A violinist plays on the G-string.

He then plays a harmonic by touching the string lightly at the same point. What is the pitch of the flageolet produced?

b. A violist plays ♩ on the C-string.
What is the flageolet produced at the same point?

c. A violist plays ♩ on the C-string.
What is the flageolet produced at the same point?

d. A violoncellist plays ♩ on the D-string.
What is the flageolet produced at the same point?

e. A violoncellist plays ♩ on the D-string.
What is the flageolet produced at the same point?

f. A violinist stops his A-string at c^1-sharp and with another finger touches the string lightly at f^1-sharp. This procedure is usually indicated by the following notation: ♩ .
What is the flageolet produced?

g. A violoncellist stops his G-string at a_1 and puts another finger lightly at b_1, thus: ♩ .
What is the flageolet produced?

8 | Stringed Keyboard Instruments

CONSTRUCTION

The idea of striking a set of strings with hammers is very old. Instruments built on this idea originated probably in Persia and from there migrated to both China and Europe. Reliefs representing such an instrument in the cathedral of Santiago de Compostela in Spain date back to 1184. To this day, one can hear the cimbalom (also called "dulcimer" and "hackbrett") played by Hungarian gypsies; and one can reportedly study it as a major instrument at the Budapest Academy. Kodály employs it in *Háry János* as an element of folklore.

The modern piano, which developed in the course of the eighteenth century, can be understood as a mechanical refinement of the cimbalom. The main difference is the assignment of one hammer to each string and the manipulation of this multitude of hammers by a lever system controlled by the keyboard. The clavichord employs a tangent instead of a hammer; the harpsichord, a plectrum. The gain in all these cases is the increase in the number of notes that can be played simultaneously and in quick succession. The price paid for this gain is the loss of direct contact between the player and the string: the mechanism that now intervenes deprives the keyboard player of the kind of intimacy with the vibrating body felt by a violinist or harpist.

The strings run across a soundboard to which they transmit their vibrations via a low bridge. We know from the monochord that the minimal diameter of a string transmits little energy to the surrounding air. The soundboard acts as a magnifier of the vibration, because its large surface can disturb a considerable mass of air. Some instrument builders have tried to "tune" the soundboard more or less to the pitch distribution of the strings above it. By thinning the soundboard under the high notes and leaving it thick under the low notes, these builders enlist a maximum assistance of resonance. You can test this quality of a soundboard by tapping it gently from side to side. Not all builders believe in this refinement. In any case, the wood used must permit the sound to travel very rapidly. If it does not, there is danger that interference (cf. pp. 65 ff.) will neutralize the vibrations of one part of the board by the contrary phases of the vibrations in another part.

The strings on a grand piano fill the range of over seven octaves, normally from $a_4 = 27\frac{1}{2}$ to $c^4 = 4224$. If string length were the only criterion, the deepest-sounding string would have to be about 153 times longer than the highest-sounding string ($4224/27\frac{1}{2}$). While varying the string length, as the mere shape of a grand piano indicates, the piano builder—as well as the builder of clavichords and harpsichords—relies equally on the other two laws of Mersenne (cf. p. 100). The lower strings become not only longer but also heavier. Thin copper wire spun around the steel string increases the weight of the string. The exact spot at which spun strings take over (usually around a_2) is critical for the builder who wants to avoid a break in the quality of the tone production. The higher strings become not only shorter and thinner but also tenser. A very short string goes easily out of tune, as baby grands inevitably prove. On the other hand, a very high tension puts a tremendous strain on the frame—up to 30 tons on the modern steel frame of a concert grand. For reasons of sonority, most tones on the piano are produced by two or three strings tuned in unison, so that this tension of 30 tons represents the combined strain of more than 200 strings.

The piano hammer is thrown at the string by the leverage extending from the key and immediately falls back after hitting the string. This is an important fact for the pianist to remember; for he is out of touch with the tone before it comes into existence. All he can do is impart a definite speed to the hammer. Afterwards he can influence the cessation of the tone. As long as he holds the key, the string is free to vibrate. At the moment he releases the key, a damper descends on the string and stops it. All dampers are also controlled by the right pedal: they are all raised at once and do not interfere with the vibrating string while the right pedal is depressed. The piano builder has to decide how hard or soft to make the head of the hammer; for the timbre of the vibrating string will be accordingly influenced. A coating of thick felt has become generally accepted. The piano builder also has to decide the spot on the string at which the hammer is to strike. The overtones originating at that spot will be eliminated—completely so if they are plucked (as on the harpsichord), and at least partially so when they are struck. Thus a hammer striking the string at one-half of its length will weaken the octave and hence all octaves above it. A hammer striking at one-third will deaden the fifth. In general, piano builders place the point of contact where least desirable overtones are located and thus dulled. This "striking distance," as it is called, lies at about one-seventh for the majority of the strings. Varying with rising pitch, it reaches about one-fourth at the extreme treble.

The shifting of the hammers by the left pedal does not change the point of impact. The left pedal moves each hammer at right angles to the string length so that the hammer strikes only two strings where there are three unison strings to each pitch, and one string (*una corda*) where there are two. The low tones, which have only one string, are also influenced by the shift of the hammer off its center. The indication *una corda*, as we can see, was precise only in the days of the early piano.

In the striking action lies the main difference between piano and clavichord. Instead of a hammer, the clavichord employs a thin metal tangent which touches and disturbs the resting string in one point and then remains in contact with the string until the key is released. As a result, the tone is much softer. The player, however, stays in constant touch with the string and can therefore influence the tone production more sensitively and steadily than on the piano.

In contrast and as a complement to the intimacy of the clavichord, the harpsichord was the brilliant concert and ensemble instrument before the invention of the modern piano. The main difference lies again in the way the vibration of the string is induced. There is no hammer, as in the piano, and no tangent, as in the clavichord. At the end of the key levers, a vertical strip of wood, called the "jack," carries a small plectrum that plucks the string. The plectrum may be a quill, a hard piece of leather, or a spine (hence the name "spinet"). In any case, the plectrum moves the string away from its position of rest to a certain point at which it passes by the string and loses any further contact with it. Because this point is always the same, fixed by the properties of the jack and the plectrum, the string always begins its vibration at the same amplitude. In short, the player cannot control the loudness by his finger action on the key. The removal of this particular shortcoming gave the successor of the harpsichord the name "pianoforte"—the instrument on which one can play soft and loud.

As a compensation for the resulting dynamic monotony, each harpsichord key can put into action more than one string. These various strings are tuned in lower and higher octaves, and also in differing timbres. By pulling a handle or pushing a pedal, the player can at will connect the keyboard with any or all of the given sets of strings. A set of strings of unified timbre is called a "stop," or "rank," or "register." To accommodate the multitude of strings, the stops are usually distributed over two keyboards.

PITCH

The player's control of pitch is limited to the keys, each of which is connected with the corresponding string (or set of unison strings) by a complex system of mechanical levers. Beyond selecting a key, a keyboard player can in no way control the pitch. In this respect, he is much worse off than any other performer. His apparent mechanical advantage quickly turns into a serious musical drawback. All other instrumentalists, let alone singers, have to work on increasing their sensitivity to the basic musical experience of pitch distinction. Singers experience pitches physiologically. String players search for the right pitch continuously with their ears and fingers. Wind players adjust by their very breath the rather approximate intonation offered by their various instruments. Timpanists have to tune and retune the membranes of their drums. Only the keyboard player can approach pitch with the detachment of a typist who counts on the appearance of a certain character as the result of a purely mechanical action. His effort is the same whether he plays high or low, wide intervals or small, consonances or dissonances. As a result, because of the very nature of his instrument, a keyboard player, more than any other performer, is in great danger of atrophying his most valuable musical instincts. To counteract this danger, he should get into the habit of singing all that is singable in a composition before attempting to play it.

While bound by the same restrictions, the clavichordist, unlike the pianist, can produce a vibrato, that is, a small pitch fluctuation. The metal tangent stretches the string slightly in response to pressure on the matching key. Fluctuation of the finger's pressure, as brought about by "wiggling" after the key has been struck, transmits itself to the string with which the player has remained connected. This vibrato is often prescribed by composers of the eighteenth century, foremost among them Carl Philipp Emanuel Bach.

The descent of the clavichord from the monochord can be seen in the role of the tangent, not only as an agitator, but also as a divider, of the string. The tangent terminates the vibrating length of the string, the remainder of the string being muted by a felt strip. The same string may accordingly be divided by more than one tangent and thus serve several keys and pitches, though obviously not simultaneously. Such clavichords are called "fretted." They are smaller and cheaper than unfretted clavichords because of the consequent reduction of the number of strings.

The harpsichordist, like any keyboard player, operates with fixed, given pitches. The presence of ranks allows a characteristic modification. A harpsichord usually contains a stop marked 4' and, if it is a large instrument, other stops marked 2' and 16'. In the terminology of organ builders, an 8-foot pipe sounds a tone of which pitch and notation coincide (cf. p. 59). Remembering what we know of wavelength, we define a 4-foot stop as one of which the tones sound one octave higher than written and played. A 2-foot stop sounds two octaves higher; and a 16-foot stop, one octave lower than the 8-foot norm. Depressing only one key, the harpsichordist can produce as many different octave pitches as there are ranks of different octave range on his particular instrument. This shifting and doubling of octaves both upward and downward is a typical feature of the harpsichord sound.

Pitch on all keyboard instruments is set by the tuner. How he does it, we shall find out in the chapter on temperament.

LOUDNESS

Loudness depends exclusively on amplitude. The amplitude of a vibrating piano string depends exclusively on the force with which the hammer strikes it. The force of the hammer depends exclusively on the speed with which the key is depressed by the finger. If the speed is too slow, there is no sound.

This fact should dissolve any illusions a pianist might have about the connection of loudness with the pressure of his fingers, or the strength of his biceps, or the grandeur of his arm motion. A finger just a small distance above the keys can play very loudly if it descends at great speed, whereas falling from a great height at the same speed it merely assumes the additional risk of missing its aim. A soft sound can be produced from any height if the speed of the impact is slow.

A pianist's crescendo is based on a particular kind of illusion, which he can make serviceable to good advantage. Unlike a singer and violinist, he cannot swell his tone. But by playing successive tones or chords each more loudly than the last, the pianist conveys to the hearer the impression that the phrase as a whole is growing in loudness—just as the eye perceives as an ascending line what in reality need be only a series of dots arranged one higher than the one before.

The right pedal does not increase the initial amplitude, but it adds the sonority of other resonant strings which are freed of the damper. Thus eventually there might be an increase of loudness, which is, however, always compensated by the natural loss of energy in the vibrating string, resulting in a shorter duration of the tone. This must be so, for the finger after depressing the key is powerless to keep feeding fresh energy to the created tone.

Loudness is the one characteristic of a single tone which a pianist can fully control. Neither pitch, as we have seen, nor timbre, as we intend to develop, can be regulated by him in any comparable degree. This restriction is severe; but the pianist who recognizes it can gain a maximum of freedom by a mastery of the infinite shades of loudness with which a tone can be made to sound.

In principle, the clavichordist faces here the same problems as the pianist. The latter's use of the pedals and the former's of the vibrato constitute individual variants which do not significantly alter the fundamental situation. Although the general level of loudness of the clavichord is considerably below that of the piano—any noise in the room is likely to overpower the clavichord sound, and duets with another performer are futile—the relative dynamic range by itself is at least equally wide. By analogy, actors watched through the wrong end of an opera glass appear very small; but within their dimension, the range of expression is as large as any. Just as the eye soon adjusts to size, so the ear quickly "tunes down" when listening to a clavichord. On the low dynamic level, the many shadings of loudness are unimpaired.

The biggest adjustment required of a pianist or clavichordist playing the harpsichord concerns the realization that the force, that is, the speed, with which he depresses a key has no bearing whatsoever on the degree of loudness. The amplitude of the vibration is fixed by the mechanism of the instrument and beyond the control of the player. Hence, a good harpsichordist will never hit a key. By depressing it steadily, he permits the plectrum to take the string along to the point of release, whereas a quick stroke is likely to damage the plectrum (and the resulting sonority) by having the string cut into it.

The harpsichordist can play on different dynamic levels by pulling his stops. Such an action is necessarily sudden. There is no possibility of a gradual crescendo and decrescendo. The sudden shifts of dynamics are characteristic of much music written in the seventeenth and eighteenth centuries. The term "block dynamics" has been used by historians and critics. The harpsichord meets the stylistic requirement of block dynamics by the sudden shifts of loudness of which it is capable. (Aesthetically the harpsichord is therefore less akin to the piano and clavichord than it is to the organ.) As one or more registers are added, the fingers notice the increased resistance offered by the greater number of plectra. The player will have to exert comparably more force to set in motion several sets of strings instead of one; but the amplitude, that is, the total loudness, is always fixed by the registration and not by his touch. Some harpsichords have a special pedal that moves the plectra farther under the strings. As a result, the amplitude of all vibrating strings is increased, because the plectrum now takes the string along to a greater distance from the position of rest. Again, block dynamics eventuates.

Because the position of the plectrum determines the amplitude of the vibrating string, this position can be regulated so that certain tones do not stick out among the others. This operation is called "voicing." It is usually accomplished by a simple turn of a screw, a function of maintenance rather than of performance.

TIMBRE

Is the pianist correct who thinks that he can vary the timbre of a single tone by his "touch"? He is wrong. All he can do is depress the key at a certain speed, thereby regulating the loudness. No pressing or wiggling can possibly influence the overtone constellation, because, even at the moment of impact, the player is totally out of touch with the string. The only variable at his disposal is the speed of his finger, hence the velocity of the hammer, hence the loudness of the tone. Any lingering illusions have been removed by experiments at the University of Pennsylvania in which single tones were first played by a famous pianist and then duplicated by a cushioned weight falling on a key. No difference could be heard. The identity can be seen in the following illustration of the recorded sound curves: the upper curve in each pair records the tone produced by the pianist's finger; the lower, by the mechanical weight.[1]

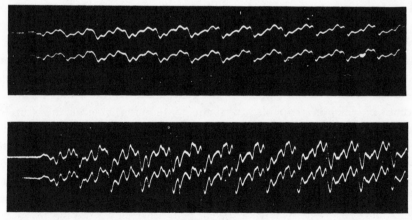

Figure 43

After this categorical denial, a few qualifications must be mentioned—so tangential that they do not fundamentally affect the truth of our initial asser-

[1] Harry C. Hart, Melville W. Fuller, and Walter S. Lusby, "A Precision Study of Piano Touch and Tone," *The Journal of the Acoustical Society of America*, VI/2 (October, 1934), pp. 80-94.

tion. First, the degree of loudness will somehow influence timbre by favoring higher overtones—in any case, by changing the overtone constellation. This kind of changed timbre, however, is a concomitant of loudness and can never be produced independently of it. Second, we cannot ignore the possible effect made by the noise of the finger falling on the key. The noise amalgamates with the nearly synchronous tone and modifies the total overtone constellation. What is commonly referred to as a "hard touch" may be explained by a strong surface noise executed at high speed with relatively little weight behind it. The situation is comparable to one in which a light hammer is used for driving a heavy nail: heat and noise and perhaps a bent nail will be the intrusive result. Third, the overtone constellation can be minimally influenced by the player at the moment the tone ceases. At the moment it begins, the player controls only the loudness. While it sounds, the player has no contact with the tone. But the speed with which he releases the key will be reflected by the manner in which the given overtones die out. A rapid fall of the damper erases the given partials differently from a slow fall. At the moment of cessation only, the form of the vibration can be modified. This is a minimal control by the player at a moment that comes too late in the life of a tone and lasts too briefly to be very significant. Good players, however, use even this minimum control to good advantage. The pianist's touch, in short, cannot influence the timbre of a single tone, but the termination of this touch can do so. Fourth, and finally, the pedals quite obviously influence the timbre. The use of the right pedal permits all strings in the piano to swing freely. The principle of resonance will create sympathetic vibrations in many of the strings in response to the one that has actually been struck. The sum of all these vibrations influences the shape of the original vibration. The new overtone constellation is heard as a new timbre. The use of the left pedal on a grand piano lets the hammer hit only two strings instead of three. A loss of loudness and a poorer overtone constellation, that is, a different timbre, are the result. On uprights, the left pedal moves the hammers, not at right angles to the strings, but closer to them. Consequently only the amplitude, that is, the loudness, is thereby regulated.

In general, then, it is safe for a pianist to remember how severely limited his control is over the timbre of a single tone. The situation changes drastically when more than one tone is played; and the pianist may find solace in the thought that any composition in his repertoire is likely to consist of a multitude of tones. Two tones may be struck simultaneously or in close succession so that the first still sounds when the second begins. In either case, the two overtone series will influence each other mutually. Resonance will amplify some overtones, interference will cancel others. Whatever the subtle interplay, a change of timbre is the result. The pianist's control over loudness indirectly affects the resultant timbre. By striking one of the keys harder, the pianist brings to life this tone's higher partials. The infinite possibilities in which the player can grade the dynamic relationship of two tones are reflected in an infinite variety of timbre shadings. The "singing legato" melody is the result of a good performer's carefully holding over one tone to the next while sensitively diversifying the successive dynamic levels. Staccato tones, which by definition are detached from their neighbors, cannot be employed for coloring and therefore usually give the impression of sounding "dry."

The timbre potentials of two tones are steeply increased when three and more tones are in play. The principle of mutual modifications of the various overtone series remains the same; the applicable combinations are boundless. Try to imagine the subtle dynamic variants just within a four-part chord. The slightest dynamic change of any tone will immediately influence the total overtone constellation, and hence the timbre, of the chord. All the more reason for a pianist to master the one aspect that lies legitimately within his grasp: complete control of varying degrees of loudness for every finger independently of similar efforts by all companion fingers.

The clavichordist must consider the same limitations and possibilities as the pianist. In addition, however, he must remember that the vibrato at his disposal influences the form of any sound curve, that is, the timbre of a particular sound. The clavichord presents the same potentials of tone production as the piano *plus* the increased sensitivity of contact with the string. This "plus" vitally affects the control over pitch and timbre. Whereas a clavichordist can thus easily transfer to the piano without particular handicap, the pianist trying to play the clavichord will have to learn several additional skills.

In the variety of timbre lies the main advantage of the harpsichord over both the clavichord and the piano. Each stop expresses a new timbre. Two stops permit three combinations; three stops, seven; four stops, fifteen; and so forth according to mathematical laws of combination. This palette of timbres comes to life in addition to some devices open to the pianist, such as the holding over of one tone into the next to influence the total overtone constellation. Moreover, the harpsichordist can play contrasting timbres at the same time by employing two keyboards. Solo melody and accompaniment, or leading voice and companion, can thus be neatly separated, and timbre made to serve the clarification of polyphony. Finally, most harpsichords have a special "lute stop," regulated by a pedal or knob. This is a damper on all strings which muffles the established registration so as to approximate the sound of a lute. The richness of timbre on a harpsichord far surpasses that of most other instruments, the organ excepted. Renouncement of this richness is the price one pays for the privilege of playing soft and loud on the pianoforte.

PIANO STYLE: AESTHETIC CONSIDERATIONS

It is thus correct to state that on the piano (1) the player can control only the loudness of the tone, (2) no actual crescendo or diminuendo is possible, (3) no true legato is possible (which would be from peak to peak). Yet we do seem to hear different timbres, we do hear crescendos and diminuendos, we do hear continuous melody.

Of all instruments, the piano is perhaps the one where the gap is widest between what happens physically and what happens as an end result in our soul. The existence of such a gap is by no means unique, peculiar to the piano, but here it is most striking and astonishing. We are never simply receivers of sense impressions, but rather we are actively engaged in the making of the final inner product. We should therefore consider sense impressions as suggestions rather than as facts. There is no essential difference between the suggestion of a graph by points and the inner creation of a melodic continuum by the "tone points" provided by the piano. Time and again, the suggestive character of acoustical impressions asserts itself. We hear, not so much what there is, but what we want to hear. Whether we hear *D*-sharp or *E*-flat depends on the context. So does our choice of interpretation between "to," "too," and "two." In general, raw facts are meaningless until they are connected by some inner principle, such as causality, recurrence, "from simple to complex," and the like. In each case, the goal of the inner transformation is a morphology or what the psychologist calls a "gestalt." The process is as unavoidable as it is irresistible. We are incapable of not seeing the imaginary line connecting isolated points. We are incapable of not hearing the melody underlying a succession of tones, even when they are played staccato. We supply the interpretation of crescendo and decrescendo while actually hearing only a succession of tones of different degrees of loudness. The suggestion of timbre as a product of changes in dynamics and in articulation (legato-staccato) has well demonstrated its irresistibility by the belief among pianists, which is still lingering on, that they are able to produce it physically—an obvious impossibilty.

In suggestive or, as we should rather say, evocative power, the piano is foremost among music instruments. It has been jokingly called an "illusion machine." The truth of this remark is borne out by the fact that a well-defined piano style—Chopin being the exception—really does not exist. The piano has not created a style of its own as have the organ, the harpsichord, and even the clavichord. Taken at face value, the piano is simply a percussion instrument. Significantly enough, it has thus been used, stripped of its evocative power, only in recent times (Stravinsky and others). At all other times, the piano was used to evoke the prevailing sound ideal of the period as manifested in the respective orchestra style. Mozart's and Beethoven's piano music represent Mozart's and Beethoven's orchestra. Liszt's piano style reflects the Wagner-Liszt orchestra as truly as Debussy's piano music suggests Debussy's orchestra. The ability to create a strong illusion also makes the piano the one instrument on which one can reproduce, with some degree of satisfaction, symphonies and operas.

When one leaves the clavichord or harpsichord to play the piano, one is at first shocked by the crudeness or even vulgarity of its naked sound. But soon, and inevitably, imagination resumes its work and recreates a world of significant sounds. The piano is the poorest and the richest of our music instruments.

Exercises

1. Play a tone or chord. Slowly depress the right pedal while holding down the keys. Listen to a slight increase in loudness and a change to a richer timbre, due to the sympathetic vibrations of higher strings responding to overtones.

2. By putting standard weights on a key, determine the weight necessary to depress a key. Repeat this experiment on pianos of different manufacture and draw conclusions as to the average weight required. Do different ranges on the same piano require different weights?

3. Place a finger on a key and, always in contact with it, depress it with maximum speed. Then decrease your speed until it falls below the minimum required to produce a tone.

9 | Wind Instruments

RELATION TO STRINGED INSTRUMENTS

A body of air is capable of vibrating because it possesses elasticity and inertia. The longer a body in relation to its diameter, the more "musical" the tone, that is, the closer to a pure manifestation of a fundamental with its harmonic overtones. A string approaches this ideal best among the vibrating bodies used in music. In this sense, the tone of a string is more musical than that of a metal stick (triangle) or of a membrane (drum). Strings and pipes emerge as the best natural producers of purely musical sounds.

As a first approximation, one may think of pipe length in terms of string length. In both strings and pipes, the vibration produces standing waves, which may be thought of as a combination of outgoing and reflected traveling waves. The velocity of a traveling wave changes with the nature of the medium. The pitch of a string or pipe, all dimensions being unchanged, will consequently fall with an increase of the inertia of the material of which the string is composed, or of the gas which fills the pipe at a given pressure. The application of this fact to stringed instruments requires heavier strings for lower pitches. The application of the same fact to pipes is of a different order, for we cannot at will use lighter or heavier air. In pipes of the same kind we must actually lengthen the air column when we wish to lower the pitch.

The temperature of the air affects the pitch of a wind instrument to some extent, because warm air weighs less than cold. As the temperature rises, the molecular motion grows faster and the density decreases. The pitch of a wind instrument accordingly rises with the temperature and falls with it. In principle, a string should behave similarly. In actuality, however, a string stretching with the heat becomes not only thinner but also looser. The lessened tension more than counteracts the lightened density. The pitch of a string therefore

122

drops as the temperature rises (because of lowered tension) and rises as the temperature falls (because of increased tension). This dissimilar behavior of strings and pipes is unfortunate in concerts when the usually rising temperature of the hall pushes them out of tune in opposite directions. In humid summer climates, however, the expansion of the wood that holds the pegs might at times exceed that of the strings, in which case the pitch rises. This phenomenon is especially noticeable in pianos, and even more so in harpsichords and clavichords. For the latter, rises up to a minor third have been observed in New York, which resulted in the eventual snapping of the affected strings.

A vibrating string communicates only a very small amount of energy to the surrounding air. In all stringed instruments, therefore, the string is made to communicate its vibrations first to a soundboard or soundbox, which in turn communicates the vibrations to a much larger volume of air outside the box. This amplification takes place at the expense of the energy of the string, the vibrations of which now abate faster because of it. Pipes do not need this kind of arrangement, because the vibrating body itself is the very air which, enclosed by a tube, communicates through an opening with the surrounding air. On a stringed instrument, the soundbox needs the string to put the larger volume of air into motion; on a wind instrument, the pipe needs a small vibrating device to set the air column vibrating. Here the analogy ends. In stringed instruments, the soundbox is the servant of the string, because the soundbox vibrates at the frequency of the string to as many different pitch demands as possible. In pipes, on the contrary, the device (such as reeds or lips) exciting the air column is the servant of the latter, because the reeds or lips adjust to the frequency of the tube, which vibrates for the most part in the modes of a definite fundamental.

In both situations we have what is called a "coupled system," but the mutual effects of the elements involved in each case differ. The communication in the coupling of the string to the outside air is poor, whereas that of the pipe to the outside air is good. The coupled system in wind instruments, moreover, consists really of three, rather than just two, elements; for the player himself, by supplying the breath, becomes part of it. He is physically coupled to the instrument in such a way that he can modify the behavior and interaction of wind supply, vibrator, and air column. The coupled system of wind instruments is thus the more complex.

In stringed instruments, the behavior of air is important only in the study of the soundbox—a very important object indeed, but still one that only influences the quality and loudness of the tone. In wind instruments, on the other hand, we are very much concerned with the behavior of air, because it plays here an essential role as the sound-producing body itself. We might very well study the relationship of string length, frequency, and pitch while completely neglecting the change of the wavelength of the string into that of air; for air acts here only as a transmitter. We cannot afford a similar neglect in the study of pipes, because now the laws of the transmitter become those of the sound source itself.

CONSTRUCTION

The air column confined by a tube is excited by a vibrating agent. According to the mode of excitation, three families of wind instruments may be distinguished: lip instruments (which are commonly referred to as brass instruments), reed instruments, and flues or flutes. In all three types, a steady air stream, which could be likened to the bow in stringed instruments, is transformed into a vibratory air motion. What accomplishes this transformation? When a steady stream is coupled to a device capable of vibration, a growing uniform disturbance results. We may think of an obstacle making the stream discontinuous—of a resulting turbulence, because the steady stream has become instable.

In brass instruments, the vibrating device forming the obstacle is constituted by the lips. They are pressed against a mouthpiece which happens to look like a small cup on trumpet and trombone, and like a small cone on the horn. The lips, capable of vibration, open and close in rapid alternation under the impact of the air stream. The resulting turbulence is the transformation of the flow of air into a vibration. In reed instruments, the function just ascribed to the lips is performed by a single or double reed. Clarinet and saxophone are characterized by the former, oboe and bassoon by the latter.

In flutes and flue pipes of the organ, finally, no mechanical device intervenes. Thus one might say that they are the only "pure" wind instruments.

The air is blown directly against an edge, on the two sides of which it alternately forms vortices and eddies. As a result of periodic rarefactions and compressions, the air flutters in the vicinity of the edge in much the same way as a flag flutters in the wind. The exact manner in which the eddies and vortices operate has not yet been satisfactorily interpreted.

One might be tempted to conclude that the length of a pipe is identical with the wavelength of the tone produced by that pipe. Remembering that the product of wavelength and frequency equals the speed of sound, 1125 feet per second, we might compute the wavelength of $a = 440$, for instance, by dividing the speed by the frequency, or $1125/440 = 2.55$ feet. We might then expect to hear the tone $a = 440$ from a pipe 2.55 feet long. This, however, is not the case. The pipe need only be half that length.

To understand this relation of pipe length to wavelength, we might profitably imagine first the situation arising in a pipe closed at both ends (although this situation does not actually occur in music practice). The behavior of the vibration of the air here resembles that of the vibration of a string. A string is fixed at both ends, where no motion can take place. The same situation holds true for a pipe stopped at both ends. The ends are nodes, that is, points of rest free from vibratory motion. The first mode of vibration in such a pipe closed at both ends could be represented graphically in this manner:

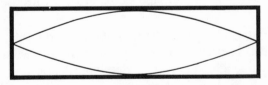

Figure 44

The standing wave is composed of an outgoing wave and its reflection. The to-and-fro together make up one complete vibration. What we see above in the pipe stopped at both ends is one-half of the wavelength; the complete vibration takes twice the length of the pipe. Although for obviously practical reasons no wind instrument embodies this type of pipe, the operation described can be demonstrated in a physics laboratory by the device called the "Kundt tube."

Let us now open one end of the pipe. Such a pipe closed at one end is called a "stopped pipe." The closed end is still a node, but the open end is now a loop:

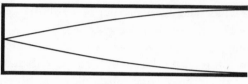

Figure 45

The pipe length is here only one-quarter of the wavelength.

Presently let us open both ends of the pipe. Such a pipe is called an "open pipe." The situation becomes similar to that created by the pipe stopped at both ends; but the nodes have become loops, and the loop is now a node:

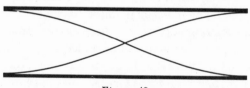

Figure 46

The pipe length, as before in the pipe closed at both ends, is one-half of the wavelength.

Both open and stopped pipes are used for music instruments; they stand side by side in an organ. To produce a pitch of a certain wavelength, an open pipe need only be half as long as the wave; and a stopped pipe need only be a quarter as long as the wave. A stopped pipe sounds consequently one octave lower than an open pipe of the same length. Organ builders make use of this principle by stopping a pipe rather than doubling its length, particularly in low registers where the pipe length threatens to grow enormous.

The graphs of the standing waves in pipes open or closed at both ends look symmetrical, whereas the graph of the standing waves in a stopped pipe looks asymmetrical. In this asymmetry lies the cause that a stopped pipe produces only every other overtone—more precisely, only the odd-numbered partials. The open pipe produces even-numbered as well as odd-numbered partials, and so would the pipe stopped at both ends if it could be heard. Stopping a pipe at one end, everything else remaining equal, consequently not only lowers the pitch one octave but also changes the timbre decisively.

All these observations are true of cylindrical pipes. The situation in a conical pipe is so complex that various explanations—proper to aerodynamics more than to music—have filled the literature.[1] We accept the fact that the closed conical pipe acts like an open cylindrical pipe, although the reasons are by no means simple. It can produce both even-numbered and odd-numbered harmonics.

The behavior of a cylindrical pipe can be easily checked by means of an ordinary cardboard mailing tube of small diameter. Let us say that we wish to produce middle c. Standard a having a frequency of 440, c below has the frequency of 440 × 3/5, or 264. The wavelength being equal to the velocity divided by the frequency, we get for c a wavelength of 1125/264, or 4.26 feet, or 4 feet 3⅛ inches. Half of this wavelength gives us the pipe length for $c =$ 25⁹⁄₁₆ inches. We cut the mailing tube to this length, tap it or blow across one end, and actually hear the desired tone. If we repeat the experiment of blowing across one end while stopping the other end by hand, we shall hear the tone one octave lower, c_1. In this simple experiment, the coincidence of the calculated and the actual lengths is fairly exact. In practice, there is often a small discrepancy because the boundary between the enclosed air and the free air is not sharply defined. The small correction that has to be applied is known to builders and players of wind instruments as "end correction."

The distinctions between open and stopped pipes, and between cylindrical and conical pipes, are manageable in theory. In the building of actual instruments, however, complications arise which have led to conflicting explanations. Thus a well-known book on orchestration, for instance, calls the flute a conical instrument, whereas an equally well-known dictionary of music maintains that the flute has a cylindrical tube. Neither statement is completely incorrect. The flute has a cylindrical bore, it is true, but the bore passes into something resembling a parabola toward the embouchure. The result is a total shape that has been called "cylindro-conical." Furthermore, the flute appears to be stopped at that same end; but the embouchure hole nearby makes this

[1] *The Physics of Music* by Alexander Wood (6th ed.; New York, 1961), p. 112, offers both a clear exposition and ample bibliography.

instrument, in effect, a pipe open at both ends. The clarinet has in principle also a cylindrical bore, but there are deviations from the pure cylinder in several places. The single reed of the clarinet acts like the stopped end of a pipe, so that in the clarinet the odd partials are prevalent. We say, "prevalent," for the even partials are not altogether absent. The oboe has a conical bore, which apparently offsets the closed-pipe effect of the double reed to such an extent that the instrument behaves like an open pipe. The mouthpieces of both the single-reed and double-reed instruments never have a cylindrical bore, and in consequence there are further variations of pitch and general behavior.

In short, whereas the first principles of vibrating air columns can be readily grasped, what actually happens in wind instruments is extremely complex. Much of it is not yet completely understood. Instrument-making is still largely empirical; and science, on the whole, has been of small help. E. G. Richardson remarked in 1929: "The makers of wind instruments and builders of organs have found their way to the present stage of evolution of their products via the thorny path of trial and error. Science has lagged so lamentably behind in applying itself to the problems involved in this trade, that at present one can do little but dot the i's and cross the t's of the manufacturer with a view to indicating the possibilities of applied physics in this direction."[2] By and large, this statement is still valid today.

PITCH

In principle, a tube sounds only the partials of its fundamental. (The occasional formation of some other tones in the pipe, for example, those favored by the submultiples of one of the overtones, or those produced by lip and reed glissandos, need not concern the musician at this point or detract from the force of the opening statement.) In some cases, the fundamental itself does not speak. In other cases, some overtones cannot be produced. But the series of tones itself remains unaffected by these or other restrictions.

The technique of playing an overtone rather than the fundamental is known as "overblowing." It is controlled by a changed embouchure and increased

[2] *Cantor Lectures on Wind Instruments from Musical and Scientific Aspects* (London, 1929), p. 27.

air pressure, but on woodwind instruments in particular it is further facilitated by a small hole bored at a certain point and often manipulated by a special "speaker key." At the location of this small hole, the air wave forms a loop. If it is bored at a half-way point of an open cylindrical pipe, for instance, the pipe sounds mainly its octave at the expense of the fundamental. The instrument, one says, "overblows at the octave."

In order to play scales on wind instruments, we have only two methods at our disposal: shortening or lengthening the pipe. The first method is applied to woodwind instruments; the second, to brass instruments. The shortening is accomplished by holes drilled in the tube. The lengthening is accomplished either by a slide (trombone) or by the coupling, through valves, of additional tubing to the main stream (modern trumpet, French horn, and tuba).

The ancestor of all woodwind instruments on which one can play a melody is the panpipe or syrinx, which consists of a number of pipes of different length, tied or glued together in the shape of a raft. If we take the longest pipe and bore fingerholes into it, we obtain a single pipe functioning as a syrinx. At the same time, we have introduced several complications. Not only are the number and position of the holes limited by the possibilities of the hand, but the size of the holes must be adapted to the fingers. If the hole is too large, it cannot be effectively stopped. As a result, the holes have a smaller diameter than the pipe, although ideally both diameters should be identical to obtain the same tone from the pipe shortened by the hole as from the corresponding separate pipe. The location of the holes has to be accordingly adjusted. The effective pipe length from the mouthpiece to each hole is somewhat greater than it would be on the separate model pipes of the syrinx. The deviation from the exact proportion varies with the size of the hole. Moreover, each hole requires its own end correction, which also varies with the size and position of the hole.

One cannot help but feel admiration for the players who through sheer skill were able for centuries to overcome the acoustical handicaps of such instruments, at least to a degree sufficient to render them musically acceptable. The deadlock in the conflict between acoustical requirements and playing possibilities was broken toward the middle of the nineteenth century by Theobald Boehm (1794-1881). Starting with the flute, the instrument on which he himself was a virtuoso, Boehm revolutionized woodwind construction. He first considered the acoustical requirements of the instrument, while temporarily ignoring the player. He then invented a mechanism of levers and keys that was designed to bridge the gap between the requirements of the remodeled instrument and of the hand.

Woodwind instruments that overblow into the octave (flute, oboe, saxophone, bassoon) obviously need at least 11 holes along the side of the pipe to shorten the wavelength, that is, raise the pitch, to every required tone within the octave. Woodwind instruments that overblow into the twelfth (clarinet) correspondingly need at least 18. Actually, all these instruments have extra openings—anywhere from an additional 3 to 15—for different reasons. First, a fast trill between a fundamental and a harmonic is impossible; hence special openings are drilled for alternative pitch controls around the break of registers. Second, the natural limitation of the human hand makes further alternatives for certain pitches both desirable and practical. Third, the lowest fundamental and often the tones near it become so conspicuous in timbre and tricky in intonation when overblown that their higher octaves (or twelfths, as the case may be) are normally assigned to special holes. Most interesting, lastly, is the fact that on most woodwind instruments the lowest tone is not identical with the original fundamental tone of the pipe. The situation is best understood from a historic angle. Since the basic proportions of most woodwind instruments were established, the length of the tube has been extended in most cases while the diameter has remained the same. In terms of the original proportion, then, the pipe is longer than it should be, and extra holes regulate the production of pitches on the extension. In practice, of course, one may rightly consider as the fundamental the lowest tone that corresponds to the total length of the instrument; for each extension creates a new, bona fide fundamental. But historically, mindful of the original proportions, one is justified in suggesting that all woodwind instruments are basically built on C or on F. This situation appears in unadulterated form throughout the recorder family. Actually the respective fundamentals of the flute and oboe are middle c; of the clarinet and English horn, f_1; and of the bassoon, f_2. On the flute, as on the recorder, the fundamental and the lowest possible tone coincide. On the oboe, English horn, and clarinet, a special hole extends the range a half-tone below the fundamental (on some oboes, there is a provision for an additional half-tone). On the bassoon, the entire span of a fifth below the fundamental is produced by the manipulation of special holes, f_2 being the lowest tone normally overblown. All these low tones which we have called "extensions" are used only occasionally if at all for the production of partials.

Keeping all these complications in mind, we may think of the main woodwind instruments in the following way for practical purposes:

The flute is a flue pipe. It acts like an open cylindrical pipe. The player can overblow into the octave and into the double octave. The tones of the first octave may be thought of as fundamentals; most tones of the second octave, as overblown 2nd partials; and the tones of the third octave, as modifications of the overblown 3rd, 4th, and 5th partials.

The oboe has a double reed on a conical pipe that acts like an open cylindrical pipe. The instrument overblows at the octave, but the player can produce the 2nd, 3rd, and 4th partials.

The clarinet, with a single reed, acts like a stopped cylindrical pipe. The player overblows into the twelfth; but besides relying on the 3rd partial in the first overblown register, he also employs the modified 5th, 7th, and 9th partials for the highest register.

These three woodwind instruments have been singled out because they represent three basic types. Any other woodwind instrument appears as a variant of one or the other type. A diversity of cone or cylinder bodies of various proportions can be mated to single or double reed mouthpieces. Each mutation is a new acoustical event. The saxophone, for example, is played with a single reed but behaves in general similarly to the oboe. The bassoon is played with a double reed, but the conical bore expands at only about half the angle of the oboe's, and the fingerholes run on a slant through as much as two inches of wood. The number of possible manifestations of a principle is infinite. There is no end to different shapes and names. They make sense when related to a norm.

In brass instruments, the first principle used for pitch variation is the harmonic series. In a secondary operation, the fundamental can be changed. This is the reverse order to pitch control on woodwinds, where the fundamental is changed before the complementary technique of overblowing is applied.

The reason for the primacy of the overtone series in brass instruments lies in the great flexibility of the vibrating lips when compared to a fixed mouthpiece or reeds. This difference can best be understood if we compare the vibratory agents with one another. A recorder, like an organ pipe with a fixed mouthpiece, needs a minimum wind velocity to speak. If the velocity increases beyond a certain degree, the pipe immediately overblows. Under these circumstances, almost no crescendo is possible. If the pipe is provided with a reed, a certain variation of loudness becomes possible. In this respect, the brass player's lips used as reeds are most successful; for the whole problem of overblowing versus crescendo rests on a delicate balance between wind velocity and lip tension. This is another one of those situations of which the player is very much aware and which he solves empirically as best he can, but about which nothing very significant and precise can as yet be learned from the physicist. In any case, the brass player not only has a wider dynamic range than the woodwind player, but with the sensitivity of his lips he can coax from the tube the harmonic series to an extent that is impossible on woodwinds, and he can influence the various pitches considerably.

The brass player's skill can be measured by his ability to "find" any prescribed partial tone by his embouchure and breath alone. A clear mental idea of the desired tone comes first. Here lies the main task for beginner and virtuoso alike. The fundamental is very difficult to produce, and so are the partials above the 16th. The Classical composers, particularly on the horn, seldom risked excursions beyond the 12th partial. The following passage from a Mozart symphony presents an extreme (K. 550, trio, m. 75-78):

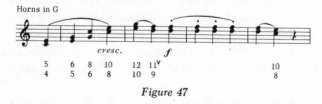

Figure 47

As a result, trumpeters and hornists playing instruments without valves have a very limited number of tones at their disposal, a fair compensation, perhaps, for the inherent technical difficulties. Moreover, the morphology of the overtone series (cf. pp. 50 f.) allows nothing but skips across the first three octaves and admits conjunct melodic phrases only above the 8th partial. This natural limitation inevitably influenced the shape of musical phrases typical of Classical scores (Beethoven, Piano Concerto No. 5, first movement, m. 49-52):

Figure 48

Bach and his contemporaries were more daring; his trumpet parts in particular demand a masterly agility (Brandenburg Concerto No. 2, first movement, m. 29-30):

Figure 49

In the three examples just given, one cannot fail to notice that certain tones do not occur in the overtone series. How can the player blow f^1 and a^1? These tones lie slightly below partials that are easily produced: f^1 just below the 11th, and a^1 just below the 14th. The player in the times of Bach, Mozart, and Beethoven "found" the adjoining higher partial and tried to lower the pitch by whatever trick he could think of: stick his hand into the bell to stop the pipe partially, or adjust the position of his lips against the mouthpiece to change the wavelength. Such devices helped utilize the extraneous 7th, 11th, and 13th partials. While admiring the ingenuity extracted from a player by a limitation, we may concede that these tones did not sound well. In any case, they sounded different from the open tones. It is worth mentioning that a brass player can also raise the pitch of a natural overtone by thrusting his hand deeply and very tightly into the bell, thus effectively shortening the tube, while using a particular embouchure. The sound effect, however, is so special that the Classical composers and players apparently avoided it.

The overtones were easy to read in our three examples because all were notated in C major. This accommodation stems from the composers and not from the authors of textbooks. Actually, the three compositions are, respectively, in G minor, E-flat major, and F major. Following the practice of their time as conditioned by the limitations of trumpet and horn, the composers prescribed "horns in G," "horns in E-flat," and "trumpet in F," as the case might be. This prescription implies the use of different instruments, contingent upon the key of a composition, each tuned to a different fundamental with all its concomitant overtones. The advantage of notating any overtone series in C, regardless of the absolute pitch, is equally obvious to the player and the student. He need master only one norm, which stands for all keys. The reader of a score, on the other hand, has to fabricate his own transposition.

The trombone, as the name reveals, is a big trumpet. The unmanageable length of a bass trumpet necessitated a winding, which already in the fifteenth century appeared in the form known to us today: the tube is cut into two U-shaped pieces one of which slides into the other. This mechanism permits the player to change the total length of the tube at will in order to alter the frequency and pitch of any vibration with nicety and precision. It also makes possible a glissando in any part of the range. The trombonist still depends on his control of the overtone series. By changing the length of the instrument, he shifts to a new fundamental, as it were, with all its accompanying partials.

Three centuries later than on the trombone, during the lifetime of Haydn, a mechanism based on a comparable idea was introduced on horn and trumpet which increased the player's control over pitch. A removable length of tubing, called "crook," was inserted between the mouthpiece and the instrument. Depending on the length of the crook, the fundamental was lowered by a halftone or whole tone, so that a player could change a horn in F into a horn in E or E-flat. Although hampered by the inconvenient exchange of crooks, a player could now produce on the same instrument the overtone series of a variety of keys.

This idea was developed and perfected in the nineteenth century, during the lifetime of Brahms. Three coiled tubes, clearly visible to the spectator, are permanently fastened to the main tube inside the main loop. Each of these coils can be joined to the main stream of air by a valve within easy reach of the player's left hand. The length of each linked coil is measured in such a way that one lowers the pitch of the whole tube by a semitone; one, by a whole tone; and one, by one and a half tones. The valves can be used singly or in combination. The maximum extension of the wavelength consequently lowers any given pitch by three whole tones. This is exactly the gap between the second and third partial tones:

Figure 50

With the help of the valves, the brass player can now play all distinct pitches that fall between the natural overtones. He finds the higher partial and then lengthens the tube as desired. A fourth valve and coil that are now often added to some brass instruments, particularly to the tuba, lower the pitch a perfect fourth. With all four valves in action, the player can actually fill in all the missing tones even within the first octave.

There is a price to pay for every improvement. Although the principle of additional tubing seems practical, it creates the new difficulty of drastically changing the proportions of the fundamental pipe. The side effects on all attributes of the tone are obvious, and the player's skill is taxed by a fresh problem.

LOUDNESS

The wind player's control of loudness is not unlike the singer's. Here as there, the air pressure is the main determinant. In the case of the wind player, the amplitude of the entire vibration, on which loudness depends, corresponds to the primary vibration of the lips or reeds, in the same manner in which in the case of the singer it corresponds to that of the vocal cords. In principle, the two lips of a brass player and the double reeds of an oboist and bassoonist approach most closely the mechanism of the two vocal cords.

The air pressure directly defines the velocity with which the player's breath escapes, and hence the amount of air that passes into the instrument within a given time unit. Theoretically, a player can increase the loudness of a tone by

increasing the velocity of the escaping breath while keeping the mouth opening constant; or by widening the mouth opening while keeping the velocity
constant; or by letting his air pressure influence both the velocity and the
opening. Practically, however, the wind player seldom varies his mouth opening when trying to increase the loudness of a tone; the risk of thereby changing
the entire character of the tone is too great. He relies primarily on increasing
the speed with which his breath agitates the lips, reeds, and total air column
of the instrument. Just as the speed of the pianist's finger represents the force
that is translated into loudness by the instrument (cf. pp. 114 f.), so the speed
of the wind player's breath assumes a comparable function.

The player's lips and the reeds, as the case may be, offer an elastic resistance
to the force of the air stream. Without this resistance, the air would never be
transformed into a tone. Let us consider the extreme cases of a maximum and
a minimum resistance. In the former, the air cannot escape at all, for the outlet
is tightly shut. In the latter, the breath escapes completely as in a person who
exhales quickly through a wide open mouth. In neither case does the air pressure produce a musical tone. For this purpose, the resistance of lips and reeds
(or of a singer's vocal cords) must be at an appropriate optimum. Neither the
air pressure alone nor the resistance alone can define loudness. In the proper
relation of the functioning of both lies the real control of dynamics.

Here the player's empirical knowledge continues to surpass by far any physical analysis that has been attempted thus far. Future research is likely to
bring us new important information. Some basic questions still beg for an answer. Why, for instance, does the clarinet have a wider dynamic range than
the oboe; and the oboe, than the flute? Why, in view of the fact that some
harmonics on the oboe are louder than the fundamental, do we still hear the
fundamental loudest? It has been said that the trumpet is louder than the
French horn because of the different shapes of the mouthpiece; the longer
tubing of the horn, moreover, offers a greater dampening resistance to the vibration than does the shorter and thicker tubing of the trumpet. The relation
of diameter to length also seems to enter as a dynamic determinant. The exact
interplay of all these agents, however, awaits a satisfactory scientific investigation—all the while that performers on wind instruments happily blow away at
any desired degree of loudness.

The relation of the speed of the escaping breath to the resistance of the lips and reeds, on which the amplitude of the resulting vibration depends, controls not only loudness but also the process of overblowing. Speaker keys or special holes merely facilitate this process without carrying the primary responsibility for it. Because the basic operation of a wind instrument is to assist in the transformation of air into tone, one need not wonder at the fact that loudness and pitch are related to the same action; for in the wind player's breath lies the formation of all the characteristics of a tone.

TIMBRE

The discrepancy between theoretic understanding and practical application is as great in regard to timbre as we have found it to be in regard to loudness; and in both cases, the musician continues to proceed empirically.

The point need not be belabored that the smallest change in any agent participating in the creation of a tone on a wind instrument evokes a change in the actually sounding overtone series and hence a corresponding modification of the total timbre. Breath, mouth, lips, reeds—these most personal agents are subject to the minutest fluctuations of the player's will. Any change of pressure or position necessarily brings about a fresh overtone constellation; and if a new pitch or a new degree of loudness is the primary purpose of such a change, timbre can justifiably be thought of as an attribute that perpetually characterizes the other properties of a tone. The singer Luisa Tetrazzini claimed that every pitch has its appropriate "natural" timbre, and many wind players would agree.

There are, however, some purely technical devices which enable a wind player to modulate his timbre. Foremost among them is the alternative between a fundamental and a harmonic, and between various harmonics. On the clarinet, for instance, d^2 is ordinarily played as a modified 5th partial of b_1-flat; but it will sound thinner and lighter when produced as the straight overblown 3rd partial of g. All wind players have similar alternatives. The technical requirements of the musical context might dictate their choice, but the most audible difference will be one of timbre.

The timbre variation between a fundamental and an overtone, particularly in the woodwinds, is so great that their respective domains are referred to as different "registers." The same distinction, but to a lesser degree, applies to tones reached through overblowing at different intervals, for example, the first and the second overblown octaves on the flute. The difference in timbre becomes progressively less noticeable the higher one ascends in the harmonic series. A main task for a wind player is to minimize the break between registers. A smooth transition between registers is equivalent to a minimal vacillation of timbre.

The player can further affect the timbre by his physical treatment of the end hole. The mere direction in which it points affects the tone color—be it the air space toward the audience, the floor below him, or the absorbing cloth of his suit. Hornists, most of all, continuously employ the right hand, which holds the end of the instrument, to muffle or open the given sound. Where the use of the hand is impractical, as in a trumpet and trombone, special mutes introduced into the end modify the timbre more than the loudness. For such a purpose, almost any object will serve—from a handkerchief to a jazz trumpeter's bowler.

The instrument builder carries a primary responsibility for the timbre of his product, although he, too, relies more on personal and traditional experience than on scientific data. A crucial factor is the relation of the diameter to the length of the pipe; for on proportion, here as in the consideration of pitch and loudness, rests the musical quality of a tone. A rich harmonic series eventuates when the diameter becomes very small as compared to the length. Among tone producers, a string is accordingly the closest materialization of this ideal. Among the brass instruments, the highest ratio, of about 1:300, is found on the trombone. Among woodwinds, the situation—as we have seen in other respects —is complex and seems to defy pinpointing. The flute, with the relatively low ratio of diameter to length of about 1:30, is the weakest in harmonics. The clarinet, with a smaller diameter than the flute, sounds richer. Beyond a certain ratio, however, a pipe risks losing the low overtones altogether so that the many high thin partials gained in the process would not add up to a musically satisfactory tone. The physical variations in the wave form that result from an alteration of the ratio between diameter and length of a pipe have yet to be investigated. The irregularities in the bore of any pipe make exact calculations and—from the builder's point of view—predictions most precarious. We understand the amazement with which Richard Strauss, in his treatise on orchestration, records the fact that a specially constructed contrabass oboe shared not at all the expected timbre of a bassoon.

An obvious modification of the basic proportion is the addition of a "bell" to the open end of the pipe. Apart from intruding on the true proportion, a bell affects the resonance and the radiation of the sound, especially when the lower holes are closed. It thus changes the timbre in more than one way. On an instrument like the English horn, moreover, the bell also—and perhaps mainly—functions as a resonant cavity, from which the instrument's lower tones gain their particular character.

The paramount role of proportion also explains why the shape of the mouthpiece on a brass instrument has a direct bearing on timbre. The old recipe, "the shallower the cup the more brilliant the tone," is an excellent generalization.[3] Accordingly we see that trumpets are fitted with cup-like, the mellower horns with funnel-shaped, mouthpieces. The respective proportions of these mouthpieces are radically different from each other. Similarly, all the dimensions of the various mouthpieces and reeds on woodwind instruments influence the tone significantly.

The connection between material and timbre is less phenomenalized than one tends to expect. The rigidity of the pipe walls seems to be as important as the resonance of the material. Yet, the metaphysical experience of materials is part of the tone experience. There is an inner relationship between the nature of the instrument and the nature of the material out of which it is fashioned. One has become accustomed to metal flutes and to metal saxophones, not to mention the metal strings on violins. It might be possible to make a perfect trumpet out of plastic, and yet we shudder at the thought if we have any feeling left for such connections.

Exercises

1. To what length must you cut an open cardboard mailing tube to produce the tones f_1, c, e, g?

2. To what size must you cut a cardboard tube closed at the lower end to produce the same tones?

3. The following phrase for a French horn in E-flat without valves occurs in

---

[3] Cf. Cecil Forsyth, *Orchestration* (London, 1914), p. 19.

the third movement of Beethoven's Ninth Symphony. Exactly how did the player in Beethoven's time produce each tone?

Figure 51

4. Berlioz suggests that a melodic phrase involving tones that are unmanageable by the natural overtones of one French horn be distributed among several horns of different fundamentals. In the following phrase, he lets a horn in C play the first half, and a horn in E-flat the second half including the upbeat. Exactly which partials does each player use?

Figure 52

10 | Organ

CONSTRUCTION

The acoustical properties of the various wind instruments are presented by the organ *in toto* as well as in isolation; for the organ not only contains a multitude of different pipes—flues and reeds, open and stopped, cylindrical and conical—but also, by holding them in a fixed situation, determines once for all the pitch, loudness, and timbre of each individual pipe. We may think of the organ—be it a small positive or a gigantic concert instrument—as a collection of pipes which can be connected to a source of compressed air and brought into action by a double system of valves. The pipes are mounted on a wind chest which supplies air under a definite pressure. In the Renaissance and early Baroque periods, this pressure lay between 40 mm and 90 mm (meas-

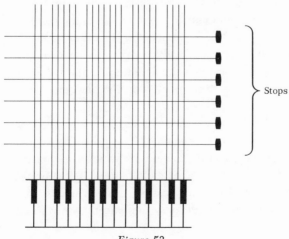

Figure 53

ured by the rise of a water column); modern organs have increased the air pressure to as much as 2000 mm. One set of valves couples to the wind chest a series of pipes unified by a similar structure, that is, a similar loudness and timbre. Such a set is called a "register," or a "rank," or a "stop." The number of ranks may vary from one to over one hundred. The other set of valves, at right angles to the first, opens up all pipes of the same pitch. It is operated by the keyboard. Both sets of valves must be opened for the pipe to sound. The stops are pulled by the player before he begins a piece, and the keys are depressed as he is playing. Each pipe of the organ can thus be understood as representing a particular tone at a particular point in a system of coordinates (see Fig. 53).

Without having to be concerned in this context with the transmission problems of the two sets of valves, we want to suggest that there is a qualitative, and not merely an engineering, difference between the two extreme possibilities of mechanical levers and electrical contacts. The former, characterizing what is known as "tracker action," preserve a desirable personal intimacy between the organist and his instrument. They favor the organist's sensibility to the resistance of the valve against being pulled away from the closure. The resulting correspondence in the movements of key and valve permit the organist to influence the speed with which he initiates and also terminates the vibration of a pipe; and we remember the decisive influence of the transient sounds at these moments on the total characteristics of a tone. With a tracker action, air flows into the pipes in a manner comparable to the controlled flow from the lungs to the glottis. Electrical contacts, on the other hand, start any vibration much more suddenly and explosively, comparable to a cough. They practically eliminate any time lag between the action of the key and the speaking of the pipe. They also admit of more freedom in the placement of the pipes relative to the keyboard.

A basic distinction exists between flue pipes and reed pipes. The former function like a flute: air blown against a sharp lip sets off the vibration. Reed pipes employ a curved elastic tongue, of which the vibration closes the air stream intermittently. The reed is called "free" if it vibrates back and forth through a tightly fitting slot, and "beating" if it hits against the frame of the slot.

A free reed cannot be significantly influenced by a resonator coupled to it.

The relatively small size of a harmonium and mouth organ, which use exclusively free reeds, is partly explained by the fact that any tubes coupled to the reeds would be musically superfluous. The tone of a free reed unfolds in a kind of rubbery crescendo, because the vibration, comparable to the movement of a pendulum, grows wider until it reaches an optimal amplitude.

A beating reed, on the other hand, is decisively influenced by a resonator coupled to it. The frequency of the reed and that of the resonating air column link most intensively if the walls of the latter are parallel to each other. In such a case (as exemplified by the krummhorn), the pipe does not speak at all if the frequencies of reed and resonator do not coincide almost exactly. Conical pipes (for instance, the trumpet reeds) are much more tolerant; and extremely shallow cups (as found, e.g., in the vox humana) have no bearing on pitch but only on timbre by the manner in which they select overtones. The attack of a beating reed is precise, without any increase of the initial amplitude of the vibration, because the movement of the tongue is abruptly braked by the frame of the slot.

PITCH

Each key controls one tone in as many different occurrences as there are ranks. For reasons of differentiation but also of practical distribution, the pipes are usually spread over several keyboards, called "manuals" for the hands and "pedals" for the feet. An organist can accordingly play on three different keyboards at the same time—a fact amply utilized by the trio-sonata literature of the seventeenth and eighteenth centuries.

The pitch of each pipe is basically determined by its length. This fact accounts for the organist's terminology. The length of an open cylindrical flue pipe sounding the low c_2 is almost exactly 8 feet. Therefore, a rank sounding as played, and thus constituting a norm, is referred to as an "8 foot," although, of course, all pipes above c_2 in that rank are proportionately shorter. The keyboard here has the potential of a transposing instrument. Lower octave ranges are accordingly referred to as "16 foot" and "32 foot"; higher octave ranges, as "4 foot," "2 foot," and "1 foot." Pipes sounding the fifth of a key adopt an analogous nomenclature: 10⅔' (= 32'/3), 5⅓' (= 16'/3), 2⅔' (= 8'/3), etc. The same principle applies to the naming of thirds (6⅖', 3⅕', etc.), sevenths 2²/₇', 1¹/₇'), and so forth. The pitch of any of these pipes is ascertained by our relating it to the 8-foot norm, for example: 2⅔ ÷ 8 = 8/3 ÷ 24/3 = 8/24 = 1/3.

In short, one might assume that the proportions developed on the monochord apply here equally well. They do so in theory, but every organ builder is aware of physically conditioned deviations, which he must correct in practice. As we remember from the wind instruments, the loop of the vibrating air column in a pipe extends beyond the opening at the end. As a result, the wavelength is slightly longer, and the pitch consequently slightly lower, than expected. The calculated length of a pipe must therefore be shortened by a certain measure to produce the desired pitch. This end correction equals empirically approximately one-third of the diameter of the pipe (varying inversely as the wind pressure) although more complicated formulas have been submitted (between $\pi/8$ and $4/3\pi$ times the diameter). The mouth of the pipe, too, demands a correction of the kind that the lips of a wind player could more readily supply than the fixed structure of an organ. The organ builder must incorporate in his calculations such factors as the diameter of the pipe and the relation of the width of the labium to the height. Moreover, the edge against which the air is blown has a tendency to generate a frequency of its own which rises with the wind pressure and, although eventually subordinating itself to the frequency of the pipe, is likely to pull up the total pitch. Here as elsewhere in music, the organ builder's experience counts for more than the mathematician's calculation. The necessary empirical knowledge explains why the art of organ-building often runs in families as a hereditary tradition.

Remembering all these necessary adjustments, we are still correct in evoking the norm according to which a pipe half as long as another will sound an octave higher. For the pitch an octave lower, the length may be doubled or the pipe may be covered at one end—in either case, the wavelength will be twice as long. Such "stopped pipes," as they are called, are singularly practical in the construction of an organ as an alternative to monstrous pipes for the deepest tones.

As still another alternative for the creation of extremely low pitches, the organ builder often resorts to combination tones (cf. p. 67). The subcontra c_4, for instance, at a frequency of 16, requires a pipe length of approximately 32 feet. Such a pipe might weigh almost half a ton. The key controlling this pitch can be connected with two pipes of which the respective frequencies differ by 16. Such is the case, for example, with the octave ($c_3 = 16 \times 2 = 32$) and the twelfth ($g_3 = 16 \times 3 = 48$). In short, the organ builder can save himself the trouble of a real 32' pipe by employing the difference tone of a 16' and a 10⅔'. By stopping these two primary pipes, he can further halve the length; for two stopped pipes of 8' and 5⅓' (c_2 and g_2), respectively, will produce the same difference tone corresponding to 32', two octave ranges below their own.

Each key on the manuals and the pedal of an organ may thus release, not only the primary pitch it identifies, but also a variety of higher and lower pitches that may be coupled to it.

In reed pipes, the length of the reed determines the pitch. Because heat and cold affect metal much less than they affect gases, the pitch of a reed pipe is less influenced by temperature changes than the pitch of a flue pipe. Organists who tune their reed pipes more frequently than their flue pipes actually adjust a more easily manageable minority group to the runaway intonation of the rest.

LOUDNESS

Because the air pressure throughout the instrument is constant, and each pipe is mounted in a fixed position, the organist cannot vary the given loudness of a tone. He may, of course, pull more or less stops and thereby determine the total volume of his particular rendition. He may contrast the dynamic level of one manual against that of the other. But once the registration has set a degree of loudness, the player can no longer influence it. Some modern organs attempt to modify this restriction by a "swell box"—a chamber around the pipes with shutters that open and close at the player's will to produce dynamic fluctuations of the given tones. The device seems extraneous to the nature of the instrument, but it has its advocates.

The loudness of a rank or of a pipe is the organ builder's problem. As in the human voice, the amplitude of the vibration depends on the air pressure and the resistance opposed to it. In practical terms, the loudness of a pipe increases with high wind pressure and a wide pipe lip. The loudness decreases with a diminished air supply, narrow lip, and small diameter. The relationship among these factors mirrors attitudes of style and taste. Organs of the nineteenth century, favoring high wind pressure, often contain similar ranks, usually distributed over several manuals, which furnish different dynamic levels of the same kind of pipe. The current trend, not unlike that of the late Renaissance, recognizes the principle (which every good singer knows to be valid) that high pressure distorts the tone quality and that a minimum of air can well create a maximum of tonal beauty. It is low wind pressure that maintains the decent tone of Italian Renaissance organs, to give an example, in spite of the prevalence of heavy lead pipes.

The restriction to block dynamics on the classical organ (i.e., sudden shifts instead of subtle gradations of loudness) is more than compensated for by the almost unlimited richness of timbre.

TIMBRE

A register spans the whole or part of the keyboard. The pipes forming a register, differing only in size, are unified by a similar structure—they are "voiced," as the technical term goes, to sound with the same loudness and timbre. One register thus presents one particular timbre. An organ has as many different timbres at its disposal as there are registers and combinations of registers. The number of possibilities, even in a medium-size organ, is staggering. Three registers yield 7 combinations (A, B, C, AB, AC, BC, ABC); ten registers, 1023; fifty, 1,125,899,906,842,624; and one hundred, a number of 31 digits.

The timbre of a pipe is conditioned by the same factors that participate in the construction of any wind instrument. Among them, the concept of *mensura*, or scaling, assumes a particular significance. *Mensura* refers to a proportion involving the diameter of a pipe. The traditional measure relates the diameter of a pipe to that of the pipe one octave lower in the same rank. Until about 1300, all pipes of the relatively small organ were built at the same width of a "pigeon's egg" (from 24 to 30 mm). The *mensura* then applied only to the pipe length, and the proportions followed those of the monochord with appropriate end corrections. But as soon as the enlarged keyboard and the trend toward a clearer timbre definition necessitated a scaling of the width of the pipes within the same rank, established geometric methods were applied to the construction. Architecture supplied the series 1, 3, 6, 10, 15, 21, 28, etc. ($= 1+2+3+4$ etc.), which yielded scalings of $3{:}6 = 1{:}2$ (length); $6{:}10 = 3{:}5$ (flue); $10{:}15 = 2{:}3$ (stopped); $15{:}21 = 5{:}7$ (reeds); $21{:}28 = 3{:}4$ (reeds); et cetera. Among these scalings, 3:5 comes closest to the ideal norm of a unified timbre throughout an entire set of flue pipes, as actually found in the rank called "diapason," the backbone of each manual. One notes with interest that this elemental *mensura* is a close approximation of the age-old "golden section."[1]

The late Baroque used other calculations which occasionally refer to a secret number—an organ builder's *arcanum*. Whatever the mathematical formula, proportion determines the timbre of a pipe as much as it determines pitch, loudness, and musical phenomena in general.

Proportion also governs the dimensions of the pipe mouth. Notwithstanding mathematical differences between various builders, there is musical agreement that the sound becomes duller and muddier the higher the mouth in relation to its width. The timbre is further controlled by the proportions of almost all parts of the pipe, even of the air channels.

A special timbre is supplied by the mixture stops. They are compounds of two to twelve ranks of narrow, open diapason pipes to each note, tuned to various upper harmonics, among which octaves, fifths, and thirds predominate. These enforced overtones add bright sharpness to each tone. The German mixtures are often arranged in double-choirs for increased iridescence. The smallest pipes cannot be tuned exactly, and their beats create a characteristic glitter.

---

[1] The "golden section" is the solution of the problem to divide a line in two parts so that the ratio of the smaller part to the larger part is the same as that of the larger part to the whole:

$$\text{minor} : \text{major} = \text{major} : (\text{minor} + \text{major}).$$

The division can, of course, be continued indefinitely. If the whole line equals 1, the major equals .618 . . . . This figure is approached by the Fibonacci Series ($1 : 2 : 3 : 5 : 8 : 13$ . . .), of which the first five terms affect the musician because they exhibit the major triad.

A similar principle underlies the mutation stops. Unlike the pure mixtures, however, they contain traditionally the fundamental below a selected set of mixed partials, and the pipe forms vary. Their tendency to imitate the timbre of known orchestra instruments explains the names of many organ stops, without letting us forget that the synthetic sound, for example, of a "cornet" rank (8' 4' 2⅔' 2' 1⅗') is a far cry from the real brass instrument. In recent years, mutation stops with high sevenths, ninths, and elevenths have created new timbre concepts on the organ.

All details of the pipe construction have a bearing on the overtone constellation. The material of a pipe may be wood or metal. The wood may be rigid ebony or resonant oak. The metal may be pure or an alloy. The shape of a pipe may be conical or cylindrical. The cone may be doubled by the connection of two funnels at their wide ends or by the insertion of one funnel into the other. The pipes, whatever their shape, may be open or closed.

Among reed pipes, broad and thin tongues yield a timbre characterized by rasping overtones, whereas slim and thick tongues project the roundness of the fundamental. In reed pipes, moreover, the overtones of the pipe proper are often discordant to the fundamental of the tongue; and although they quickly disappear, their initial presence lends the attack of certain reed registers a singular sharpness.

A few examples of different pipes are illustrated below:

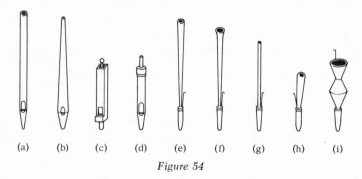

(a)   (b)   (c)   (d)   (e)   (f)   (g)   (h)   (i)

Figure 54

The open cylindrical diapason (a) favors the octaves in the complete overtone series. The conical ending of the Gemshorn (b) produces more complex higher partials and turns the timbre slightly nasal. If the pipe is covered, as in the Wood Gedackt (c), the presence of only the odd-numbered partials makes the sound hollow and dark. An interesting variant of this type is the Rohr Flute (d), where a small funnel in the stopper admits also the even-numbered partials. The funnel is usually inharmonic to the main pipe, and the prominence of these inharmonics at the moment of the incipient vibration lends the Rohr Flute its individuality. Reed pipes are decisively characterized by the shape of the resonator. A conical resonator renders the unimpaired overtone series and accounts for the festive tone of the Trumpet (e) and Schalmei (f). The cylindrical resonator of the Krummhorn (g) suppresses all even-numbered partials. The Regals all depend on the varying shapes of the resonator, which may aspire to reproduce the timbre of a human voice in the Vox Humana (h) or of a forceful roar in the Bear Pipe (i).

Exercises

1. An organ stop marked $10\frac{2}{3}$' is used together with:
    (a) an 8' stop. Why is the resulting sound undesirable?
    (b) a 16' stop. What is the resulting difference tone?
2. A mixture stop is labeled $6\frac{2}{5}$'. What tone do you expect to hear when depressing f-sharp?
3. A mixture stop is labeled $1\frac{7}{9}$'. What tone do you expect to hear when depressing b_1-flat?
4. If you have access to an organ, try combinations of various registers. Some will blend, others will remain distinct. Try to isolate and understand the acoustical reasons in each case.

11 | Percussion Instruments

CONSTRUCTION

Compared to stringed and wind instruments, percussion instruments are less sharply defined in regard both to the sounds they produce and the construction they demand. Even the modern classification spreads percussion instruments over the two categories of membranophones and idiophones. To the former belong all drum-like instruments; to the latter, all self-sounding bodies such as rods and plates (the term "plate," against its etymology, is here used to comprise tubes and bells, which can be imagined as bent plates). Notwithstanding the enormous variety of percussion instruments—from sticks beaten together by primitive man to bells cast by specialized experts—they all share an acoustical characteristic, namely, the noticeable admixture of noise-generating over tone-generating matter. Whatever the noise factors in stringed and wind instruments, they never prevail as they do in percussion instruments. The manner of playing contributes to this characteristic; for even when the pitches are relatively pure, as in a xylophone, the noise of the hammer holds its own and is aesthetically relevant.

The construction of percussion instruments persistently favors the noise element. Inharmonic partials prevail, that is, the frequencies of the free vibrations do not stand in ratios of integers to one another. The closer these inharmonics to the fundamental, the less definite the tone and the more pronounced the noise. We remember the basic rule of physics according to which the tone becomes purer, that is, the partials lie at harmonically more exact ratios, the smaller the diameter of a vibrating body in relation to its length. With this rule in mind, one need only imagine the transitional stages between a string and a rod to appreciate the qualitative differences of construction. Round

150

plates like cymbals, of which the diameter has become identical with the length, are farthest removed from a pure tone.

Yet, proportion and order govern the construction of percussion instruments as they govern everything musical. Countless forms seem possible. One might expect irregular and fancy shapes to express the irregular vibrations of noise. Actually, however, regular and simple geometric shapes prevail. Drumheads, cymbals, tambourine, and gong manifest the circle. The triangle identifies the geometric archetype by its name; moreover, it is equilateral. Orchestra chimes and the resonators of the glockenspiel materialize the cylinder. The bars of the glockenspiel and of the xylophone outline a rectangle.

PITCH

The general pitch level of all percussion instruments can be deduced by an analogy with Mersenne's laws (cf. p. 100). Although the French mathematician formulated his laws in regard to the vibrating string, we may apply the underlying principles to all vibrating bodies. As the size or weight of a percussion instrument increases, the frequency decreases proportionately. Heightened tension raises the pitch. The larger of a pair of kettledrums—to give just one illustration—is the lower; and the timpanist can raise the pitch of each of his instruments by tightening the screws that stretch the membrane.

This explanation of the relative pitches of percussion instruments does not account for the strange phenomenon of our hearing timpani and bells, in particular, one octave lower than they actually sound. If our ear mistakes the absolute octave position of these instruments, the accompanying noises doubtless add to the confusion; but the reason for the error has not been sufficiently investigated and is still unknown. As a practical application, one renders a timpani passage on the piano (as in a four-hand arrangement of a symphony) by doubling it at the lower octave; otherwise the evocation of the desired sound fails. This psychological discrepancy seems to be paralleled by the apparent discrepancy between the absolute pitch of a bell and the mass of metal involved. To produce the tone d^1-sharp, for instance,

Figure 55

the lightest possible bell will yet weigh about 220 lbs. The increase in weight is staggering as the pitch is lowered. The diameter of a bell approximately doubles with the lowering of the pitch by an octave, whereas the weight increases more than ninefold. Hence a bell sounding one octave lower, d-sharp, weighs over 2000 lbs; and a bell producing g_1,

Figure 56

nearly 10,000 lb. These enormous weights preclude the use of regular church bells in the orchestra. Wagner in *Parsifal* requires a c_2. An actual bell of that pitch, it has been computed, would weigh 409,600 lbs. Allegedly, this was the approximate weight of the largest bell ever founded, the Tsar Kolokol in the Kremlin, which was destroyed by fire in 1737 after a short existence of only three years.

Pitch has been used to classify percussion instruments, for practical purposes, by introducing a distinction between those with definite pitch and those with indefinite pitch. The dividing line, however, cannot be neatly drawn. The admixture of inharmonic partials and of noise is often so pronounced in instruments tuned to a definite pitch that our ear sometimes tolerates deviations from the prescribed pitch in percussion instruments that it would never accept in stringed and wind instruments. Thus a kettledrum may occasionally play a note out of harmony with the rest of the orchestra without being noticeably disturbing. Richard Strauss, in his edition of Berlioz' *Traité d'instrumentation*, calls attention to such an occurrence in the first finale of Verdi's *Un ballo in maschera* (between rehearsal numbers 80-81). Strauss adds that the procedure is not quite to his taste, but he concedes that the kettledrum pitch sounds too indefinite to interfere with the massive chords of the full ensemble. Inversely, the gong, which has no definite pitch, can play tricks when some component of its sound combines with an inharmonic partial of the orchestra sound and thus becomes so prominent as to give rise to a disturbing dissonance.

Because the sound of all percussion instruments contains a large share of inharmonic partials, the definiteness of pitch (i.e., the "musicality" of the sound) depends on how far removed from the fundamental these inharmonics lie. The higher up they are, the more quickly they fade than the more powerful fundamental, thus permitting the tone to become definite. The nearer they are to the fundamental, the more they obscure the exact pitch and approach noise. This condition is utilized when other instruments are employed to imitate percussion instruments. Puccini in *Tosca* (Act II, between rehearsal numbers 50-51) evokes the sound of a snare drum by major seconds in the orchestra (clarinets, bassoons, and trumpets, in turn). Mussorgsky imitates bells by a similar interval in the coronation scene of *Boris Godunov*:

Figure 57

A pianist may do likewise, for instance:

(for a bell on c_1)

Figure 58

The musician's interest in the acoustics of instruments may justifiably be most intense for the specifically "musical" ones, that is, those capable of producing definite tones; and it may gradually diminish as his studies approach the noise makers, that is, the mere rhythm instruments. A musician would learn little of value and use to him by investigating the acoustics of, say, the triangle.

Bells seem to present a special case, perhaps because their sound from churches, temples, and towers touches the lives of people all over the world. A bell may be thought of as a circular plate bent down from the center. When a round plate vibrates, nodal lines are likely to form in concentric circles and along some diameters. On a bell, one or two circular node lines run parallel to the rim; but there may be 4, 6, 8, 10, or 12 meridians from the crown down to the rim. We hear a multitude of tones. The pitch of a bell is usually identified by that of the "strike tone," which results from the impact of the clapper on the soundbow. Strangely enough, the frequency of the strike tone has thus far defied physical isolation and definition. The strike tone is so strong and bright as to drown out everything else at the moment of striking, but it fades away rapidly. In these particulars, the bell is acoustically differentiated from the gong, which acts in exactly the opposite manner. Once set into vibration, the body of a bell produces an individual set of "hum tones," a kind of inner harmony. The third, fifth, and octave predominate, often with octave doublings. These more or less predictable overtones are joined by a number of other intervals, both consonant and dissonant, varying with the size and shape of the bell. Ideally, the pitch of the strike tone should coincide with that of the "prime tone" or "main resonance tone" of the inner harmony, a term generally applied to the first overtone of the lowest hum tone. Practically, the strike tone often lies a major second, third, or even fourth above it. A minor second must be avoided to protect the sound against muddiness. An inharmonic relationship of strike and hum tones will also cause acoustical impairment if the two overtone series lie close to each other.

A good bell caster should be able to determine the pitch of a new bell in advance of trying it out—an art that has become ever more neglected. A small adjustment of the finished bell may be made by scraping the inside in order to lower the pitch. An attempt to raise the pitch of a cast bell, for example, by removing metal from the rim, only tends to destroy the basic proportions and hence the character of the particular instrument.

Bells destined for carillons differ from church bells by requiring a definite, clearly recognizable pitch. The art of making carillon bells was developed in the Netherlands, particularly by the Flemish, and reached a peak in the seventeenth century with the Hemony family in Amsterdam. The carillon belongs to the class of keyboard instruments: the bells are suspended in a fixed position, and the clappers are activated by a mechanism controlled by a keyboard. The sensuous appeal of carillons notwithstanding, melodies in the traditional Western sense rendered by bells are difficult to hear and even to recognize. They sound as if played on a gigantic piano with all dampers off. Javanese music, which is noteworthy for the highly developed participation of chimes, disks, and gongs, seems to have found a proper style arising from a deeply adequate feeling for the acoustical properties of bells and related percussion instruments.

LOUDNESS

All percussion instruments, by definition, share the manner in which the vibration of the body is initiated. Something—a stick, rod, clapper, hammer, or any practical object—strikes the instrument. The potential amplitude of the vibration is predetermined by the elasticity and resonance of the given instrument. The actual amplitude which determines the loudness of any one tone is directly related to the force of the blow. From the player's point of view, only the speed with which he hits the instrument affects the force. The expert timpanist need not raise his sticks high in order to play fortissimo.

TIMBRE

To the extent to which percussion instruments become incapable of producing pure tones, they justify their existence by the creation of a rich palette of sound colors. The complexity of the sound of percussion instruments increases with their musical impurity. Two familiar rules account for this fact. The tone will be less pure the larger the sounding body and the more its dimensions become equal to each other. As long as rods and plates—be they of glass, metal, or wood—remain relatively small, thin, long, and narrow, they produce musically acceptable tones. The celesta and the xylophone illustrate this point. But already the metal tubes used for orchestra chimes, especially the larger ones, are far from providing pure or even unmistakably recognizable tones. The problematic history of the deep *Parsifal* bells, even in Bayreuth, offers ample documentation.

The degree of purity and the elasticity of the material influence the timbre here as in other instruments. Whereas the string on a violin, for instance, has to be spun so neatly that every part of it vibrates qualitatively like every other part, the metal rod of a triangle, according to its function, vibrates so impurely that the noise far exceeds the tone. In either case, the material must possess enough elasticity to vibrate at all. Empirical tradition seems to have supplied many standards. The composition of bell metal, for instance, which is an alloy of four parts of copper to one part of tin, has hardly changed over the last thousand years.

Sources of noise are less limited than sources of tone. Richard Strauss, in his opera *Elektra* (245-246) prescribes the scratching of a cymbal by a triangle stick. Gustav Mahler, in the finale of his Sixth Symphony (129 and 140) has the percussionist hit a dull object with a hammer. Arnold Schönberg, in the *Gurrelieder* (Part III, off and on between 9 and 29; and in the three measures before 65) employs rattling iron chains ("einige grosse eiserne Ketten").

The percussion player himself can influence the timbre by choosing the material, whenever he is free to do so. The heads of his drumsticks may range from hard wood, which is appropriate to some Baroque music, to lush felt, which is more suited to Romantic compositions. Moreover, he chooses the place at which he strikes his instrument, thus varying the overtone constellations. Hit nearer the rim, a drum generally favors higher partials; nearer the center, lower partials. At the point of impact, in any case, the particular overtone corresponding to the nodal division of the point is weakened or eliminated.

The behavior of membranes has much in common with that of plates. Classic experiments with plates were initiated by E.F.F. Chladni, who was born in the same year as Mozart (1756) and died in the same year as Beethoven (1827). Chladni's experiments have a strong aesthetic appeal. He made visible the distribution of the standing waves in a plate by first covering the plate with a tenuous layer of fine sand and then generating vibrations in the plate, held in a vise, by tapping or, better, bowing. The plate may be circular or of any other regular shape. The sand is shaken away from all vibrating areas and arranges itself in interesting and beautiful figures along the nodal points and lines of the vibration pattern. Different sand designs represent different overtone constellations (Fig. 59).

Figure 59

When diameters and circles combine, they give rise to complicated designs and in general to many inharmonic tones. The same plate may vibrate in a variety of modes, depending on the point of fixation in the vise and method of excitation. The emergence of a definite tone can be favored if the plate is supported at points belonging to the nodal lines of that tone.

Similar experiments can be performed with sanded membranes, but here we must use resonance as the exciting force. Stretch a very thin sheet of rubber over one end of a tube fixed in an upright position. Strew fine sand on the membrane. If you now sing or play a loud tone nearby to which the tube is resonant, the sand will arrange itself in a Chladni figure. In this setup, you can change the fundamental by modifying the tension of the membrane, or the diameter of the tube (and hence also of the membrane), or the length of the tube. The air enclosed by the tube acts as a resonator. The situation is reproduced in the kettledrum, where the kettle takes the role of the tube.

Analyses of the complex sound of a bell are made by means of resonance. Adjustable tuning forks are set in vibration and tried on various points of the bell. The ear nevertheless remains the supreme judge, for the fork analysis reveals only the presence or absence of partials and not their dynamics and timbre. All-important for the composition of the bell sound is the "rib," a name specifying the profile or one-half of the cross section of the bell body. A study of the ribs of various bells is the basis of any further knowledge about them. The thickness of the rib is one of the factors determining the formation of overtones. The illustration on the next page shows the rib of a bell by the "Stradivari of bell casters," the Fleming Gerhard Van Wou (1480-1520), with the places where the first 3 partials originate; the thickness at the soundbow is assumed to be 1.

The total sound of a bell is, of course, strongly affected by the relation of the strike tone to the hum tones; but the number, pitch distribution, and loudness of the latter most characteristically define the timbre. In a good bell, the hum persists for a long time with relatively little dynamic loss. Every new strike refreshes it. In this regard, a bell differs from a xylophone, of which the partials are weak and fade quickly. The main determinants among the bell partials are the lowest three, that is, the prime tone together with its lower

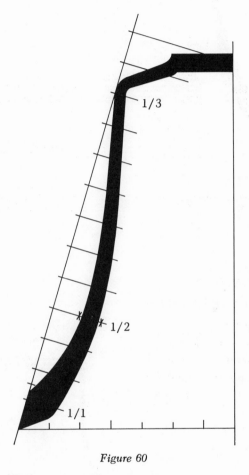

Figure 60

octave and upper fifth. For yet unexplained reasons, a minor third above the prime tone often becomes characteristically audible. The higher partials are strongly dissonant. Before the evolvement of the Gothic form, which has remained standard to this day, inharmonic partials prevailed.

Following are sketches of some bells with relevant information:

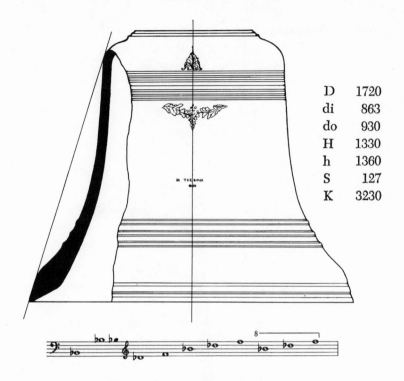

D	1720
di	863
do	930
H	1330
h	1360
S	127
K	3230

ABBREVIATIONS:
(all distances in millimeters)

D	Diameter at lip
di	Inner diameter at shoulder
do	Outer diameter at shoulder
H	Height measured on tangent between lip and shoulder
h	Inner height
S	Thickness of rib at soundbow
K	Weight in kilograms
●	Strike tone
O	Hum tones

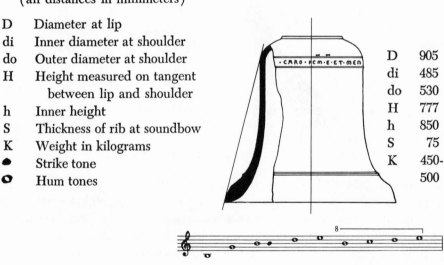

D	905
di	485
do	530
H	777
h	850
S	75
K	450-500

Figure

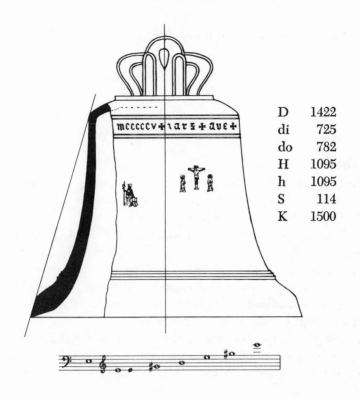

D	1422
di	725
do	782
H	1095
h	1095
S	114
K	1500

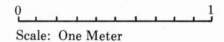

0 |_____| 1

Scale: One Meter

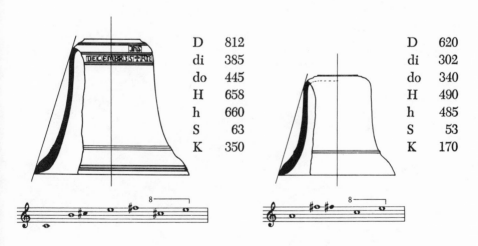

D	812
di	385
do	445
H	658
h	660
S	63
K	350

D	620
di	302
do	340
H	490
h	485
S	53
K	170

Exercises

1. Strike a rod or plate. Experiment with the following variables:
   (a) Strike with objects of different materials, weights, and shapes.
   (b) Change the points at which you strike.
   (c) Change the points at which you hold the struck object betweeen two fingers.

In each case, notice the varying proportion between tone and noise; the emergence or disappearance of overtones; and the manner in which the sound is enhanced or muffled according to the presence of a wave maximum or minimum, respectively, at the striking point.

2. Repeat these experiments with the struck object laid across an open box. Changing the position of the object relative to the rim corresponds to changing the suspension points. Notice the influence of the box acting as a soundboard.

12 | On the Function of the Ear

Perceptions are sense transformations of stimuli. The question has been asked whether a tone exists if nobody can hear it. The question is absurd, because what exists in that case is not a tone but a mechanical happening "out there": vibrating air.

Tone is born inside us and exists nowhere else. To speak of tone as something existing independently of our soul is a fallacy.

We are, of course, interested in learning about the ways and means of such a remarkable transformation. The anatomy of the ear is known in considerable detail. More or less adequate illustrations and descriptions of it are found in almost any book dealing with sound, and anyone wishing to delve more deeply into the subject may turn to anatomical and physiological publications. Here we need only briefly recall the principal features of the anatomy of the ear.

One distinguishes three parts: the external, the middle, and the internal (or inner) ear. The external ear consists of the visible wave collector and of a tube, which is closed at its inner end by the drum membrane. The middle ear contains a lever mechanism (the ossicles), which enlarges the vibrations and communicates them to the oval window, also closed by a membrane, that marks the beginning of the inner ear. The middle ear has a direct connection with the outer air through the Eustachian tube. The inner ear is filled with liquids. Here we find the organ of hearing proper, also called the "organ of Corti" after the Italian anatomist Marchese Alfonso Corti (1822-1876) who discovered it. It is a highly complex apparatus inside the cochlea, thus called because it is shaped like the coil of a snail shell. This is the place where the transformation of mechanical impulses into nerve impulses is accomplished before these impulses are transmitted to the brain by the auditory nerve. The inner ear also holds the organ that controls balance and direction, the semicircular canals.

On the way from the sound generator to the acoustic nerve fibers, transverse and longitudinal waves are transformed into each other many times. Whatever the original wave motion—transverse in a string, longitudinal in a pipe—air vibrates only longitudinally. The longitudinal vibration of the air is changed in the eardrum into transverse vibration, and transmitted through the lever system of the ossicles (transverse) and the membrane closing the oval window (transverse) to the cochlear fluid (longitudinal). The fibers of the organ of Corti vibrate again transversally. Finally, the mechanical vibration is transformed into nerve impulses.

Authorities agree on how the mechanism operates until it reaches the inner ear. What happens in the cochlea, however, is still far from being clearly understood. Georg Von Békésy, who is considered one of the greatest living authorities in the field, has this to say: "The words 'theories of hearing' as commonly used are misleading. We know little about the functioning of the auditory nerve and even less about the auditory cortex, and most of the theories of hearing do not make any statements about their functioning. Theories of hearing are usually concerned only with answering the question, how does the ear discriminate pitch? We must know how the vibrations produced by a sound are distributed along the length of the basilar membrane before we can understand how pitch is discriminated, and therefore theories of hearing are basically theories concerning the vibratory pattern of the basilar membrane and the sense organs attached to it.

"The problem under discussion is a purely mechanical one, and it may well seem, at least to the layman, that it can easily be solved by looking at the vibratory patterns in the cochlea. Unfortunately, this direct approach proves difficult, for without stroboscopic illumination and other special devices, it is hardly possible to observe any vibration in the nearly transparent gelatinous mass in the cochlea of a living organism."[1]

Von Békésy's experiments with models of the cochlea point toward extremely complex concurrences of various types of response of the organ of Corti (resonance, traveling waves, standing waves, telephone theory) and of neural discharges or "volleys." The accumulated knowledge has at present

[1] *Experiments in Hearing* (New York, 1960), p. 539.

not crystallized into a comprehensive, consistent, and generally accepted theory, fit to be presented in a book on acoustics for musicians. We may concede that one day the whole process, including the chemico-electrical actions in the brain, will be bared. The tone, however, will never be found. It is not an object, to be found in the outer world; and the organ of Corti, the nerves, and the brain are all part of the outer world. One might as well expect to find the soul by dissecting the body.

We emphasize that phenomena belonging to different realms cannot be explained by each other. They can only be correlated and perhaps connected to a unifying phenomenon from a third realm. Such a phenomenon is number, which has served us so well as a *tertium comparationis* between the quantitative vibration and the qualitative tone. We have stated that the influence of the inner phenomenon (tone) on the outer one (vibration) through the unifying middle term (number) is essentially of a morphological nature. Now, in nature the inner morphology tends to mold the outer morphology, and this truth holds peculiarly for living organisms and their organs. Hence we are inclined to consider the visible *morphé* as the phenomenalization of the number *morphé*.[2]

Let us illustrate what we mean by the organ of balance, the semicircular canals. There are three of them. Each has the shape of an arch, and the three arches are clustered in an orthogonal system, like the coordinates of a three-dimensional space. This is startling. We move around carrying space coordinates inside us. How is this surprising fact to be interpreted? Some people might say that apparently the organ of balance and direction is modeled after space. This is a questionable view, for three-dimensional space is not the only possible space. Moreover, we find it difficult to imagine space causing organic coordinates to grow. In any case, the assumption seems no more mysterious (nor less so!), and a lot more plausible, that an inner norm selected three-dimensional space from among many possibilities and caused the corresponding organ to grow. Consequently, this particular space becomes of particular value to us; it becomes symbolic for all other kinds of space.

The sense of space is an inner awareness. Our attitude toward it differs

2 The concept *morphé* indicates the totality of form and structure.

significantly from that most of us assume toward the outer senses. We do not generally doubt the reality of the inner space perception. We are convinced that something "out there" corresponds to the something "in here." Space exists. When it comes to the data of the outer senses, however, we display a strong tendency to consider the phenomenal world an illusion—experienced through a veil, like that of the Indian goddess Maya. Light, color, heat and cold, bitter and sweet, noise and tone—perhaps all these are vibratory happenings "out there," and perhaps they are but that. The awesome question of reality arises. Possibly our senses are cheaters, but it is difficult to admit it. It is difficult to assume that reality should be all on one side. It is difficult to concede that the world should be all mechanical. Indeed, it is rather absurd to do so. The postulate underlying this book is that there is as much reality to be found in the sense data as in those of the outer world; and, furthermore, that without the inner norms the data gathered in the outer world would be meaningless.

The ear, we submit, is modeled after the inner norms of tone. Tone, inversely, is not a chance result of the ear's constitution. This postulate should have some influence on the study of the organ of hearing, not so much on the research itself as on the interpretation of the findings. Kayser has shown that the very shape of the cochlea corresponds to a certain transformation of the Pythagorean table, the "tone spiral." Thus an observable form suddenly assumes significance through an interpretation "from inside out."

Modern physiological research has been generally determined by the questions concerning pitch, loudness, and timbre, many of which are still unanswered. Meanwhile, specific problems like that of consonance and dissonance have been handed over to the psychologists, who more often than not are trying to reach results by methods based on behaviorism and statistics. To the musician, this kind of technique and the conclusions it might proffer are meaningless. We are not at all interested in statistics telling us how many persons of indeterminate ability can distinguish an octave from a fifth. We know that the world is imperfect; there is no need for trying to prove it. We know that there are degrees in the purity of the manifestations of norms. Noise is the rule, musical tone is the exception.

The ear, like everything else, is also imperfect, although it is an amazingly sensitive and precise organ in many respects. The ear does not determine what is musically important. *We* determine *how* we want to hear what is given. Musical hearing is an act of selection. A norm helps us select the intended pitch among the multitude proposed by a vibrato. Thanks to a norm we are capable of ignoring all sorts of disturbances, such as beats, combination tones, and the pervading dissonances of a well-tempered keyboard. Without the assumption of a norm, the overtone series does not explain the major triad. Above all, the minor triad is totally inexplicable on physical grounds.

In science, it is important that the right questions be asked if the experiment is to speak significantly. The questions are directed at nature by a mind bent upon learning how nature operates. In musical acoustics, as in all special fields, the questions are more narrowly defined. We are interested, not simply in learning how nature operates, but to what extent nature selects in respect to our norms. For physiological acoustics to gain musical meaning, it should apply itself to discovering whether, and to what degree, the structure and functioning of the ear favor musically important norms. In view of the present limited state of knowledge, such investigations may be a long way off. Their results will not change music, but they will enrich our understanding of harmonical embodiments in nature.

Exercise

Study the model of an ear, as it is usually available in a department of biology, psychology, or medicine. Take the model apart, acquaint yourself with the appearance of the separate members, and then reassemble the model correctly.

13 | Acoustics of Halls

MUSICAL CONSIDERATIONS

Rhythm and pitch are the fundamental elements of music. With only these elements in mind, one can create a work of music and fix it on paper. There actually exists such a body of music from various centuries, including our own. Ensemble pieces of the Renaissance provide a typical example. Music thus conceived is always essentially structural. It exists solely by virtue of its rhythmic-metric, melodic, and harmonic structure. It may be perfectly appreciated by the act of reading (provided, of course, the reader is able to hear with his "inner ear" what he is reading). In this communication between the work and the subject, no outward physical manifestation is present. But as soon as the work is to be physically manifested—in other words, as soon as a performance is planned—a definite "incarnation" must be chosen. The musical structure now becomes an object for "instrumentation." Thereby the quasi-abstract musical thought is assigned a definite individual body. Volume and timbre are introduced. An atmosphere is created without which no living thing can exist. Now at last the music may be performed, and another stage of incarnation is reached.

The presence of favorable acoustical conditions now becomes an important element. They should be such as to allow the "musical body" to shine in its own particular light. They should enhance its peculiar atmosphere. This does not mean, as all too many people would have it today, that a condition suffices in which the music is perceived distinctly, clearly, and with crispness. The desirable acoustical atmosphere depends on the style of the music. Alpenhorn tunes, which are created in mountainous regions, are slow-moving, harmonic rather than melodic, and rely on an echo effect. Notre Dame music, written for a highly reverberant Gothic cathedral, becomes unbearable when transferred to a crisp hall. Music by Mozart, on the other hand, should be heard with clarity. Wagner and Liszt again ask for a good deal of acoustical halo. In general, halls destined for music should never be so dry as halls destined for spoken communication, although even here a certain degree of lively acoustics cannot be dispensed with. Anechoic chambers—without echoes and reverberations—are unfit for any sort of communication, musical or verbal. We shall discover later that good acoustics is not only and not mainly a function of reverberation but of something called the "resonance" of a room.

We have used the term "acoustical atmosphere." This term actually describes a psycho-physical complex; and although reverberation and resonance are of prime importance in this complex, other factors enter as well. Consider, for instance, the size of a hall. A natural relationship seems to exist, in more than one respect, between size and acoustical atmosphere. On the physical side, everything else being equal, a large hall is appropriately more resonant than a small hall; and one has indeed noticed greater subjective tolerance toward lively acoustics in the former. On the musical side, the transplanting of chamber music into a huge hall appears as a heresy, even if the acoustics were adjusted. In the Salle Pleyel, a very large hall in Paris, you can find a few seats at the farthest distance from the stage where you hear a string quartet or a solo violin as perfectly as if the sound source were only a few yards away. Nevertheless, the impression can be unpleasant and disquieting. The reason may be partly the discrepancy between the optical distance, which is immense, and the suggested aural distance, which is very small. (Incidentally, orchestra music never reaches adequate power in this hall—a phenomenon characteristic of many "doctored" halls.) But there is another reason for that particular discomfort. Intimate conversations and confessions are for the few; orations and sermons, for the many. One might similarly suggest some sort of parallelism between, on one hand, the size of the hall and of the performing body and, on the other hand, the number of people sharing the experience. The attraction of large crowds by famous chamber-music ensembles is no proof to the contrary. In such a case, as in many others, the fame of the players coupled with the concomitant advertising is more than probably responsible for the apparent reversal of a normal situation.

It is, of course, neither possible nor necessary to build as many different halls as there are acoustical requirements, be it in regard to different sizes of the performing body or to varied musical styles. The proper use of very small and very large halls is easily defined. But in between, there exist many multi-purpose halls. Compromises are as inevitable as they are possible. Actually, orchestral music, or choral music with orchestra, is rarely performed in an undersized hall. The obvious physical and acoustical limits of such an undertaking are quickly reached and apparent to everybody. The contrary situation, however, of a solo recital or a chamber-music performance in a somewhat too ample hall is often encountered. In this case, the musical interpretation undergoes certain modifications and distortions, not unlike those to which works of the visual arts are subjected when perspective is taken into account. The level of loudness may have to be adjusted. Certain delicate nuances may have to be exaggerated. The tempo, too, might be influenced. Larger halls will induce the performer to take slightly slower tempi in fast movements, whereas in slow movements the physical tone will sometimes not be quite so weighty as the composer imagined it so that a less ponderous tempo will suggest itself.

Another element of "acoustical atmosphere" is the cohesion of the public. There are halls which seem to draw people together. There are other halls in which one feels isolated among the crowd. Several reasons may concur in producing the impression of cohesion. An important one is sympathy among the listeners. It presupposes a common level of interest and understanding for the music. In our fragmented society, this occurrence is rare, to say the least. Optical factors also play a role. In some halls, parts of the audience, especially on balconies, are optically cut off from the rest—so much so that the individual cannot even sense whether the hall is filled or gaping. Finally, a hall with the proper resonance will accomplish a maximum toward promoting a feeling of audience cohesion. In such halls, one feels enveloped by the sound as if one were sitting inside the soundbox of some gigantic instrument. This final analogy sums up most of what has been said thus far. A hall is indeed a soundbox if it is well conceived. Thereby it is part and parcel of the performing body—a veritable extension of it. Hence the public will feel drawn into the musical event. It will partake of the performance. The presence of an acoustical atmosphere, through its several agents, will increase the bond between performers and audience. In this kind of ideal hall, the ideal audience is indeed, not an onlooker or "onlistener," but rather a participant.

ACOUSTICAL FACTORS

In trying to separate the various components of what is called the "acoustics" of a hall, we can profitably begin by considering the differences that exist between the behavior of sound in an open and in an enclosed space.

In an open space, sound is propagated from the source in all directions, which means that it is propagated in the shape of an expanding sphere. The paths of the sound are the radii of the sphere. The loudness diminishes with the growing distance from the source. According to only the geometry of propagation, the intensity, that is, the amount of energy striking the eardrum, should decrease as the square of the distance. Practically, this general law suffers at least three modifications. First, there are internal losses of energy in the medium, with the result that the attenuation of the intensity is more rapid than predicted by the geometry of propagation. Second, variations of loudness (i.e., of the subjective impression) do not parallel variations of intensity (i.e., of the energy at the sound source). The Weber-Fechner law states that the relation of sense responses to stimuli is logarithmic (cf. p. 61). We have noted earlier that the only case to which this law applies exactly is the relation of frequency to pitch. Yet, the law is valid for the given situation though only in an approximate manner. The result of this modification is a slower decrease in loudness than in intensity. Third, the decrease in loudness is not the same for every pitch. Indeed, the sensitivity of the ear for intensity variation is not the same in all regions of pitch.

Thus things are already rather complex, and yet we have assumed a simplified situation never to be encountered. We never listen to, or make, music somewhere in midair, but down here on earth, where at once an additional complication arises from the reflection of the sound waves by the ground we stand on. In general, this situation should result in a reinforcement of intensity and hence of loudness. Reflection, however, is never perfect. Depending on the nature of the ground—earth, stone, wood, rugs—a greater or lesser part of the sound energy is destroyed by absorption. The harder the reflecting surface, the more sound is reflected. The softer and spongier the surface, the greater the absorption. At any rate, the waves are reflected from the ground only once before traveling out into space together with that portion of the wave front that did not hit the ground.

The situation changes again as we proceed to enclose gradually the space around us. Let us first assume a wall of sufficient size, and let the source of sound be fairly near it. Little or nothing will be heard behind the wall, but the sound will be considerably reinforced for people in front of it. If we move away a certain distance and produce a shout or short tone, we shall be able to hear the reflected tone as an echo. The ear, behaving in this respect as the eye does, cannot separate consecutive impressions that are less than about 1/17 of a second apart (a number well known to producers of motion pictures). In 1/17 of a second, sound travels about 20 m, or 66', in air. A shout or short tone may be assumed to cover roughly the same time span. In order to get an echo after no less than 1/17 of a second, the sound source should be at least 20 m, or 66', from the wall. The echo becomes more distinct as the distance increases.

Let us now assume a second wall erected opposite the first one at a distance, say, of 40 m, or 132', and let the source of sound be near the first wall. Waves striking the wall at an angle will after an initial reflection be directed toward either the sides, the sky, or the ground and thus be immediately or eventually lost, as the case may be. The latter fate is shared by any wave parallel to the ground which strikes the opposite wall with only a small deviation from the perpendicular. Any wave, however, at right angles to the walls will be reflected back and forth between them.

The musician located near the first wall will hear a confused echo from the repeated reflections of the parallel and near-parallel waves. He will get the impression that the sound is being prolonged. This phenomenon is called "reverberation." It lends the primary sound a quality of liveliness; but if it exceeds a certain duration, it becomes disturbing even for music. Acousticians have attempted to ascertain the best reverberation time for concert halls. They have suggested figures lying in the neighborhood of 1.5 seconds for large halls filled to capacity. Because the reverberation time, unlike many other acoustical factors, can be easily measured, acoustical engineers in recent years have laid too much stress on it alone to the undue neglect of other factors. If we now complete our enclosure by adding the two missing walls, making the width of the hall, let us suppose, 20 m, or 66', and by covering the structure with a flat roof 25 m above ground, the reflection patterns become much more complicated. If the enclosing surfaces are hard and smooth, our hall will have a tremendous reverberation. The hall will be vastly "over-acoustical"; for some locations, the reverberation is likely to take on the nature of an echo. The drawing below shows some of the reflection patterns in our hall, and the places where an echo is likely to develop.

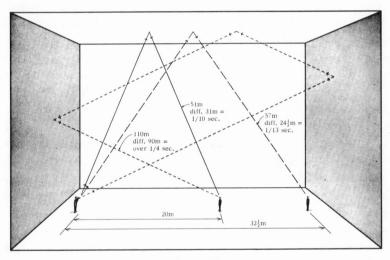

Figure 62

Drawings like this one belong to what has been called "ray acoustics" or "geometrical acoustics" in analogy to "geometrical optics." Although useful as a first step in acoustical prediction, the importance of such a drawing has often been overrated. For our hall, the prediction contained in the drawing might be fairly accurate. Such a hall, however, does not exist. It has no windows or openings for air-conditioning. It has no doors. The walls are totally bare. It is unfurnished. People are presumed not to be present. All such elements will alter the acoustics decisively. They are sound-absorbing agents. They might also scatter the sound if physical conditions prevail that favor what is known as "diffraction."

The phenomenon of diffraction is perhaps most easily explained by a comparison with the phenomenon of shadow. Both arise from the encounter of a wave with an obstacle. The crucial factor is the relationship between the length of the wave and the dimensions of the obstacle. An obstacle large in respect to the wavelength casts a shadow. An obstacle small in respect to the wavelength causes diffraction. Light waves are so small that they are stopped by an opaque object of any but miscroscopic size; a pin casts a shadow as well as a wall. By contrast, the incomparably longer sound waves are not completely stopped even by a wall. We cannot see around corners, because at the edge the minute light waves either continue straight on or are completely stopped. (Any occurring diffraction is negligible.) We can, however, hear around corners, because the edge creates a new wave center by diffracting the rather long sound waves in all directions. If there is a hole in the obstacle, light goes through in straight lines. Sound not only goes through but, as a result of the diffracting effect of the rim, immediately spreads on the other side. If you gradually close a door to an adjoining room in which music is made, the initial decrease in loudness is minimal. Only at the last moment before the door is completely shut, the loudness decreases steeply because diffraction becomes suddenly eliminated. You can notice a similar abrupt effect by plugging the keyhole in an otherwise closed door. Yet we cannot flatly state that sound casts no shadow, for some sound shadows do occur. Obstacles that are small in respect to long sound waves may be large in respect to shorter sound waves. When a marching band approaches through side streets, you hear first the bass drum. At that moment, you are in a sound shadow for all other sounds emitted by the band, that is, for the shorter sound waves of the higher instruments. The high tones become again inaudible before the low tones after the band has passed you and disappeared around corners. For similar reasons, the low pitch of traffic noise is the one heard from a distance of a few blocks.

Diffraction is of considerable influence on the acoustics of a hall. For all practical purposes, asperities inside an auditorium are small compared to the wavelengths of musical sounds. Ridges, railings, openings, boxes, stucco decorations—any of these may cause diffraction and disperse the sound. The presence of diffracting objects renders a hall less vulnerable to undesirable echo effects, although it may sometimes also cause disturbing interference. In general, the beneficial effects seem to prevail.

Whatever the contribution of the factors discussed thus far, we consider the resonance of a hall of paramount importance. As applied to architectural acoustics, the term comprises the phenomena of both free and forced vibrations induced in the enclosed air space. Although resonance, like all sound phenomena, is ultimately a quality of the air surrounding our ears, yet we may distinguish between the resonances set up directly in the air space and those indirectly transmitted to it. The first kind is called "volume resonance"; the second, "panel resonance." By volume resonance we understand the modes of free vibration of the enclosed air. By panel resonance we understand the soundboard effect of the room enclosures.

At the present time, resonance is the most neglected among the acoustical factors. Hastings does not mention it at all.[1] Jeans admits that rooms seem to have a "characteristic pitch," but he explains it as that pitch for which the absorption of the room is a minimum and therefore the reverberation time a maximum. He adds that to suppose the pitch in question to be one of the free vibrations of the room is "untenable for innumerable reasons."[2] Knudsen, however, states: "The subject of resonance as applied to architectural interiors deserves more attention than it has been accorded by physicists and acoustical engineers. The subject is at present chaotic and mostly empirical, and yet it must be admitted that resonance is an important factor in the acoustics of music rooms."[3] This remark is regrettably as valid today as when it was printed over thirty years ago.

The existence of volume resonance can be ascertained easily by a familiar experiment. Sing a chromatic scale in a tiled shower stall or a small tiled bathroom. An impressively strong resonance will become audible at a definite low

---

[1] Russell B. Hastings, *The Physics of Sound* (Saint Paul, 1960), chapter "Architectural Acoustics."

[2] James Jeans, *Science and Music* (Cambridge, 1947), p. 215.

[3] Vern O. Knudsen, *Architectural Acoustics* (New York, 1932), p. 56.

pitch within the range of a man's voice. If one dimension of the room is much larger than the other two, as is likely to be the case in a shower, the enclosed space acts like a pipe stopped at both ends. The frequency of the resonance is that of the fundamental; it can be readily computed and checked by experiment. What actually happens in this situation is that one of the free vibrations of the "pipe" is excited by the singing. In rooms of which the dimensions are less different from each other, the resonant frequencies of all three dimensions have to be taken into account. Knudsen has developed a formula for calculating to any desired degree of completeness the modes of free vibrations in rectangular rooms of any given size. His experiments were conducted in a small room, 8' : 8' : 9.5'. Of the results he says: "The author's experiments on volume resonance show that all the natural frequencies predicted by the classical theory of the vibration of the air in a room are not only present, but are so prominent as to affect profoundly the quality of sound in small rooms."[4] And further: "The reverberation of sound in a small room consists (at least primarily) of the free damped vibrations of the air in the room. . . . Existing theories of reverberation are incomplete, and may lead to gross errors, especially for frequencies which have wavelengths comparable with the dimensions of the room."[5] Here it is clearly stated that geometrical acoustics cannot fully account even for reverberation, and that wave theory should be favored in all advanced and really important acoustical problems.

Knudsen himself calls his investigations of more than thirty years ago "preliminary," and so they are. One might expect that such an important change in approach to room acoustics would have brought on a rash of fresh experimental data and theoretical elaborations. Nothing of the sort has happened. By and large, architectural acoustics is still thought of in terms of reverberation and absorption of sound.

The neglect even today of resonance compared to reverberation can probably be laid to the account of the pioneers of scientific acoustics concerning halls. Professor W. C. Sabine, of Harvard University, above all, had approached the subject from the side of ray acoustics because this method prom-

[4] Ibid., p. 414.

[5] "Resonance in Small Rooms," The Journal of the Acoustical Society of America, IV (July, 1932), 36.

ised to be more immediately amenable to theoretical simplification and to measurement. His followers had all the success that could be hoped for from such a limited approach, but their very success seems to have blurred the outlook. Evidence against the panacea of the reflection-absorption theory has meanwhile been accumulating. There have been some spectacular failures among recent concert buildings which amply evidence the "incompleteness of existing theories of reverberation." Diverse musical-acoustical properties of existing halls showing nearly the same reverberation time should have been interpreted as a warning. The Musikvereinssaal and the Konzerthaussaal in Vienna have a measured reverberation time of, respectively, 1.35 and 1.6 seconds for the tone $c^1 = 512$, with a capacity audience present. On the strength of this comparison, the Konzerthaussaal should be the better of the two halls, but the contrary is true. The Musikvereinssaal is justly renowned for its acoustics, whereas the Konzerthaussaal is the terror of musicians and listeners alike. Moreover, once a hall is erected and found wanting, it can never be doctored by the subsequent manipulation of reverberation. At best, the crudest faults such as echoes or the production of sound foci may be minimized or eliminated. The positive gain will be small, and the hall will forever remain unsatisfactory. In short, a hall has to be "well born." In the design of such a hall, the science of reflection and absorption occupies a relatively narrow sector of the required knowledge.

If volume resonance has been neglected by science, panel resonance has been practically ignored; beyond a restatement of empirical evidence, almost nothing can be reported. In general, authors agree that wood paneling is best, because wood has the widest range of response. It has a relatively high absorption for low frequencies and a low absorption for high frequencies, in contrast to porous materials which act the other way around. Wood has long been respected as an ideal material for music rooms. In the Leipzig Gewandhaus, famous for good acoustics, the stage is connected to the wood paneling by large wooden beams, to the effect that, as Jeans puts it, "the walls of the building are made to act as a huge sound-board."[6] A wooden floor and wooden wall panels not only favor acoustical conditions in general but also

[6] *Op. cit.*, p. 216.

make halls less critical by almost automatically correcting imperfections that may have crept into the design.

Thus far our attention has been directed mainly toward the acoustical behavior of air in enclosed spaces, and much less toward the architectural structures themselves. Both are, of course, intimately connected. A good architect, however, is not merely the servant of acoustical laws. He can manipulate them to a certain degree; and the deeper his knowledge and insights are, the freer he is to do so. Thereby a new set of acoustically relevant factors appears.

Like any piece of architecture, a hall is defined by shape and by proportion.

In the term "shape," we include the inner as well as the outer contours of the auditorium. Balconies, for instance, are a part of the total shape and may play an influential role in the acoustics of a hall. But of prime importance are the main boundaries, the outer contours.

Acoustically, shapes may be divided into two classes: those that are on the safe, favorable side; and those that are on the risky, unfavorable side. To the former belong the classic rectangular shapes with flat roofs. To the latter belong all those shapes that exhibit curved surfaces, which are likely to produce sound foci and echoes. Shapes that by their very definition contain a focus or foci, for example, parabolas, should be eliminated altogether. Knudsen says: "A felicitous shape is a requirement of the highest priority. Unfortunately many architects believe that faulty shapes can be corrected by covering the offending surfaces with highly absorptive materials and by adjusting the reverberation time. Thus deluded, they adopt a fashionable construction method, such as the concrete shell, and produce a building that is an acoustical perversion. A bad shape is a permanent liability."[7]

Between these extreme failures and the acoustically safe shapes, there exist many intermediate solutions which under certain circumstances produce good results. Think, for instance, of the opera houses as they began to be built from around 1600 on, continuing through the nineteenth century. Almost all express the principle of the semicircle, though with many variations. The inherent

[7] "Architectural Acoustics," *Scientific American*, CCIX/5 (November, 1963), 88 f.

dangers of the semicircle are minimized or eliminated in these halls, mainly through the following features:

(a)   The semicircle does not present a smooth surface but is broken up by the deep recesses of boxes and balconies. These act as sound dispersers. They also doubtless influence the volume resonance, in a way which has not yet been investigated.

(b)   The first opera houses had flat or near-flat ceilings. Whenever these developed into domes, problems arose, because sound is likely to get trapped in a dome or projected into a focus. Coffering of the dome surface minimizes such undesirable effects.

(c)   Styles up to the emergence of contemporary architecture called for much stucco decoration. The sound-dispersing effect of such decorations is considerable.

It is apparent that modern structures which exhibit the geometrical design "in the nude," as it were, are much more sensitive to the slightest fault in design than were those older halls with organized and decorated boundaries. One is amused to observe a comeback of stucco decoration in the shape of those suspended sound dispersers, referred to as "clouds," which are so fashionable today and which are so apt to produce laboratory acoustics by impairing or destroying the living acoustical factor: resonance. As has been repeatedly pointed out, geometrical acoustics, of which suspended dispersers are the latest offspring, has negative rather than positive virtues. It is very useful, even indispensable, as long as it is kept in its place. It becomes harmful as soon as one tries to submit all problems to its method.

Quite logically, the question of proportion has been all but ignored by a scientific method based on ray acoustics. The concept lies outside the line of thinking of geometrical acousticians. The question of proportion, on the other hand, is very much a part of the vocabulary of wave theorists. There are some small indications that proportion, which has played an important role in the past, will be recognized anew.

The essential connection of volume resonance with proportion is clear enough. In a room resembling an actual pipe, that is, a room which is small and of which one dimension is much larger than the others, one fundamental is strong and lies within the audible range. Let now the dimensions become less unequal, and at the same time let the room grow larger. Two things will happen. First, the predominance of one fundamental over the two others will be weakened, unless the dimensions become equal, in which case all three dimensions will have the same fundamental. Second, the fundamentals together with a number of low overtones, as the case may be, will sink into the infra-acoustic region and become inaudible. Therein lies actually an advantage; for the lower the fundamental, the higher the modes of vibration that fall within the audible range. High overtones lie near to each other, and as a result the resonance grows more uniform. There are no peaks of favored pitches. The acoustician would say that the graph of the response becomes "flat"—a circumstance happening precisely where it is needed, namely, in the audible range. The change from the shower stall to the concert hall may thus be thought of as one from the organ pipe to the soundbox.

All we can add is the regrettable comment that no scientific information is available on the influence of proportion and of the variation of proportion on hall acoustics. The issue has thus remained wholly empirical. Empiricism being part of history, our whole evidence must be gathered from the past. Further thoughts on proportion will therefore be included in the following, historical section of this chapter.

HISTORICAL SURVEY

As far as we know, the history of acoustics begins with the Greeks. The gap of over two thousand years that separates us from their culture is tangibly bridged by monuments still standing and still used for performances. More or less well-preserved open-air theaters can be found in Greece of today, in what was once Magna Graecia (Sicily and Southern Italy), and in various locations of the Roman Empire (the pupil of Greece in architectural as in many other matters). The acoustics of these theaters is excellent, and we should therefore be curious to learn something about the theoretical principles underlying the design of those superb structures. Presumably, we might extract a theory from an acoustical analysis of the buildings themselves, but this first method has never been thoroughly applied. We might also look for documentary evidence, distinguishing between documents that treat of the nature and propagation of sound and others that deal with architectural acoustics in a practical way. The first are all but summed up by Aristotle and Porphyry. For the second, we have to rely on Vitruvius.

The Greeks assumed that sound was propagated in the shape of expanding cylinders, and that the angle of ascension was 45°. This opinion can no longer be upheld, for we know well that sounds expands in the shape of a sphere. If the underlying theory of the Greeks was wrong, how could they possibly obtain such splendid results? The question has not been asked, let alone scientifically investigated. We suggest that the Greek idea represents an approximation toward what actually happens in a practical situation. A glance at the following diagram shows that the expanding half-sphere centered on the sound source might well be stylized into an expanding cylinder.

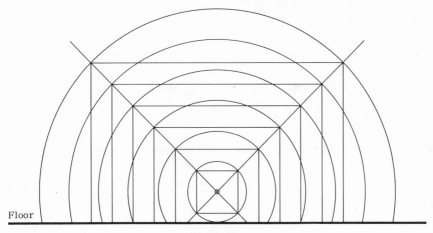

Figure 63

Perhaps there is more to it. In an open, round auditorium such as a Greek theater, waves can be reflected toward the listeners only from the floor. Among the possible reflection angles, all of which lie between 0° and 90°, that of 90° may be singled out for two properties. First, it is a limiting value. Second, it is acoustically peculiarly efficient, because the perpendicular reflection from the floor reinforces the direct waves in a manner dependent upon the height of the sound source above the floor and upon the wavelength. In principle, the higher the source elevation, the greater the formation of reinforcement lines and the weaker the effect. The same situation develops as the wavelengths decrease in relation to the source elevation. Let the elevation equal the wavelength, and three lines of reinforcement become apparent: one horizontal along the floor, one vertical from the sound source up, and one at an ascension angle of close to 30°.

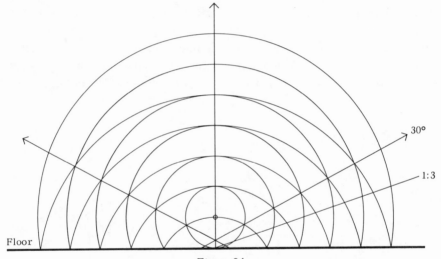

Figure 64

For a tone in the middle range of the male voice, the wavelength is slightly below six feet (1.77 m for f_1).This length is approximately the same as the elevation of the sound source, that is, of a man's head, above the ground. The slope of a Greek theater has been calculated to follow generally the ratio 1:3. If we think of the 30° line as a reference line for a zone of optimum audibility within the 45° rise of the expanding cylinder, the 1:3 rise of the steps makes a good deal of sense. Be that as it may, the subject is certainly worthy of further research.

The Greeks, Vitruvius says, not only "perfected the ascending rows of seats in theatres from their investigation of the ascending voice, and, by means of the canonical theory of the mathematicians and that of the musicians, endeavoured to make every voice uttered on the stage come with greater clearness and sweetness to the ears of the audience."[8] Often they also put to good use the phenomena of reflection and resonance. At Orange in the French Provence, for instance, a huge wall closes the rear of the stage. The arrangement proves remarkably effective. A mere whisper from the stage can be heard at the farthest seat, whereas transmission of sound in the opposite direction is negligible even in the empty theater. According to Aristoxenos and Vitruvius, resonators were built into the theaters. They consisted of a great number of vessels made either of bronze, or of earthenware when bronze was found too expensive. They were tuned to the notes of the several tone systems and were placed in niches under the seats in an inclined position with their openings down and toward the outside. At some time, the existence of these resonators had been doubted. Recently, however, such vessels were actually found in at least two places. We cannot, of course, now assess the effectiveness of the "echeia," as Vitruvius calls them. The fact that he devotes a whole chapter to their description may afford some measure of their importance at the time. Knudsen remarks: "The use of these resonators in the classical open-air theatres suggests certain possibilities of improving the acoustical quality of music rooms, and it will be seen . . . that large resonant areas of wood paneling or plaster on lath are to be found in concert halls of the highest repute."[9] Incidentally, Vitruvius says that wooden theaters need not be furnished with echeia, but that their use is necessary in theaters made of nonresonant material such as stone.[10]

After Vitruvius in the first century B.C., there is silence on the subject of architectural acoustics for nearly one and a half thousand years. Meanwhile,

[8] Vitruvius, *The Ten Books on Architecture*, trans. Morris H. Morgan (Cambridge, 1926), Bk. V, chap. iii, p. 139.

[9] *Architectural Acoustics*, p. 489.

[10] An article in *The New York Times* of 24 May 1964 reports of two projects submitted to improve the wanting acoustical quality of the Royal Festival Hall in London. One was based on the proposal to raise the ceiling by fifteen feet. The other—which was finally adopted and executed—concerned electronic resonators tuned in steps of three cycles from 300 Hz down to 70 Hz and fitted permanently into the ceiling. Thus the first suggestion concerns a change of the proportions of the hall; and the other, an application of Vitruvius' echeia.

the Byzantine, Romanesque, and Gothic styles had come and gone, leaving their extraordinary monuments for future generations to behold and admire. Acoustics does not seem to have been of much, if any, concern to the builders of the innumerable churches and cathedrals of the Middle Ages. Or, rather, we should say that acoustics was not a separate specific concern to those architects. They built according to certain principles and precepts which were believed to guarantee not only beauty but also sturdiness. We may safely assume that if they ever thought of acoustics, they would also have taken it for granted as one of the by-products of their building canons. In point of fact, many of the great cathedrals are acoustically surprisingly good, despite their enormous volume. Others are not so good; and a few are acoustically disastrous, among them Notre Dame in Paris. It is, however, not fair to judge the acoustics of cathedrals by the standards valid for works of later centuries. As remarked earlier in this chapter, the style of what is known as Notre Dame music indeed agrees with the acoustical conditions of Notre Dame Cathedral and of the other great Gothic churches, much more so than with those of ordinary concert halls.

But what were those principles and precepts we hinted at a moment ago? They were essentially Pythagorean, as they had been in Greek times, and of a kind applicable to the most varied styles. Each style creates its own world of forms and its own symbolism of forms. But all truly great periods have one doctrine in common: the doctrine of proportions. The eminent Renaissance architect Leon Battista Alberti, who wrote the first treatise on architecture after Vitruvius, defines the doctrine ably and beautifully in these words: "The Rule of these Proportions is best gathered from those Things in which we find Nature herself to be most compleat and admirable; and indeed I am every Day more and more convinced of the Truth of Pythagoras' Saying that Nature is sure to act consistently, and with a constant Analogy in all her Operations: From whence I conclude, that *the same Numbers, by means of which the Agreement of Sounds affects our Ears with Delight, are the very same which please our Eyes and our Mind.*"[11] The latter part of Alberti's statement, which we have italicized, is an old Pythagorean saying. It asserts that visual proportions are translations of musical intervals and chords, the

[11] *Ten Books on Architecture*, trans. James Leoni (London, 1955), Bk. IX, chap. v, pp. 196 f.

connecting element being number. Thus musical norms, judged by the ear and translated into space by way of the ratios appearing in the vibrating string, have presided over the choice of basic proportions in all great periods of architecture, from the Greeks to the Renaissance and beyond.

Nobody will uphold the view that the nineteenth century was in any sense a period of great architecture. Yet certain traditions were still operative, not in an actually creative way, but rather as a concomitant of a conservative and historically conscious mind. So it happened that the old doctrine of proportions was remembered and used—an empirical guarantee of visual beauty and good acoustics. It was in the nineteenth century that most of the large concert halls were built. Before the French Revolution, concert life as we know it today did not exist. One went to churches and opera houses to hear music performed by large ensembles. Public concerts were a rare exception. After the Revolution, however, "the people" gradually began to take over the role of patrons of music formerly held by the clergy and the aristocracy. Hence the need for public concert halls arose and developed.

By stating that something of the old Pythagorean proportion doctrine was carried into the nineteenth century and applied to concert halls, we do not imply that such an occurrence was the case everywhere and every time. But even as short and incomplete a list of some of the very best halls, like the one included in this chapter (cf. Fig. 65, p. 186) shows an unmistakable trend toward harmonical proportions in these buildings. We clearly sense that the architects were concerned with proportion, and we notice that a Pythagorean attitude determined their choice among an infinity of possible ratios. A glance at the list shows that the proportions are seldom unblemished by deviations, and the objection might be raised that one is not justified in applying even small corrections in order to make the measures fit what we suppose to be the intended proportions. Although valid in other areas, such an objection does not apply to morphology nor to any other normative concept. The contention that 1,001 equals roughly 1,000 is morphologically valid. It does not work when applied, for instance, to a telephone number or to a lottery ticket. Norm always implies tolerance, although there is a difficulty in saying just how much of it. Both 80 and 81 are felt to be the interval of a third. Intonation is based on tolerance. An overtone played on a wind instrument may be forced upward before jumping to the next harmonic. Even applied mechanics knows

CONCERT HALLS

[The measures given in this table are taken from Schiess, after Kayser, for Basel, Bern, Leipzig, and Zürich; from Kayser for Berlin; and from direct correspondence with the respective administrations of the Boston, Chicago, Cleveland, and Vienna halls. The three sets of numbers for each item indicate, in order, the actual measurements in meters, the assumed norms, and the proportions.]

Hall	Length	Width	Height	Hall	Length	Width	Height
Basel	35.50	21.00	14.50	Cleveland	51.83	25.61	15.85
G. Musiksaal	35.00	21.00	14.00	Severance Hall	51.83	25.92	15.55
	5	3	2		10	5	3
Berlin	32.38	12.85	9.81	Leipzig	38.00	19.00	14.30
Singakademie	32.70	13.08	9.81	Gewandhaus	38.00	19.00	14.25
	10	4	3		8	4	3
Bern	22.00	11.00	7.30	Wien	51.00	19.00	18.00
Conservatory[1]	21.90	10.95	7.30	G.Musikvereinssaal	50.88	19.08	18.02
	6	3	2		48	18	17
Boston	46.34	22.86	19.81	Zürich	30.00	12.00	9.00
Symphony Hall[2]	45.72	22.86	19.05	Kl. Tonhallesaal[4]	30.00	12.00	9.00
	12	6	5		10	4	3
Chicago	30.49	20.22	12.65				
Mandel Hall[3]	30.36	20.24	12.65				
	12	8	5				

---

[1] Original designs,	28.00	11.20	8.40		21.30	12.80	8.50
which for reasons	28.00	11.20	8.40		21.30	12.78	8.52
of building techni-	10	4	3		5	3	2
calities could not be used:							

[2] The length of the hall minus the stage is 37.80 m. Further significant proportions emerge:

Overall length	Length minus stage	Width	Height
46.34	37.80	22.86	19.81
45.72	38.10	22.86	19.05
12	10	6	5

[3] The length is measured to the back of the proscenium arch. The stage adds another 7.62 m.

[4] Later shortened by 2.50 m and thereby said to have lost much of its acoustical excellence.

Figure 65

the concept of tolerance. Similarly, proportions need not, and might not, be embodied exactly. Yet we recognize the intention through the imperfect embodiment, and we are fully justified in doing so.

Of the nine halls listed in the preceding table (Fig. 65), the proportions of three halls are based on fifths and octaves: the Musikvereinssaal in Vienna, the Gewandhaus in Leipzig, and the Conservatory Hall in Bern. The six remaining halls include the interval of the third beside the fifth and octave. The Kleiner Tonhallesaal in Zürich is the only one among the nine to exhibit the intended proportions perfectly. Unfortunately it suffered, some years ago, a shortening of its length by one-twelfth; and it has thereby reportedly lost much of its former acoustical excellence. The Conservatory Hall in Bern is of peculiar interest inasmuch as we definitely know the quoted proportions to be the intended ones. The hall, built in the present century, was designed by the acoustician Ernst Schiess, who had previously submitted two other designs which he considered preferable to the one finally accepted. Nevertheless, the hall as it stands now is acoustically faultless. Before the destruction of Berlin during the Second World War, the Singakademie had the reputation of being superior to all other halls in the former German capital. As to the Gewandhaus in Leipzig, we have already had the opportunity of mentioning the particularly felicitous use of wood. Small wonder that the resulting quality of a soundbox coupled with perfect proportioning has produced a hall of acoustical fame. The Grosser Musiksaal in Basel, although little known outside Switzerland, is in fact one of the best halls anywhere. It is as typical a nineteenth-century building as is Symphony Hall in Boston; but it is somewhat smaller and acoustically even more satisfactory.

Symphony Hall has been rightly held up as a model among American concert halls, and the proportions deserve the main credit. Similar to it, Severance Hall in Cleveland favors the third and fifth to the exclusion of the fundamental. The emphasis on the fifth in Boston as against the third in Cleveland may give favorable distinction to the former; but both halls are praised for clarity and brilliance associated with harmonic rather than fundamental sounds. Mandel Hall at the University of Chicago exhibits a special lesson in proportion. The dimensions given in our table approximate excellence, which is experienced only when the reflecting curtain at the back of the proscenium arch is lowered. As soon as the stage becomes part of the hall, the total length embodies the major seventh (15); and proportions and acoustics both suffer.

We have postponed to the end a discussion of the Musikvereinssaal in Vienna, a hall of high repute and of puzzling measures. It is the most elongated of the listed halls, 8:3. Also, there is the unusual module of 1.06 m, of which the three corrected dimensions are multiples. We further notice that the height is nearly equal to the width. Finally, and principally, this height represents the 17th harmonic, a nonsenaric value. Could the tolerance extend to as much as one meter? Might we be justified in supposing the height equal to the width? We cannot give the answer at the present time, we can only raise the questions. Future developments and refinements in the techniques of investigating volume resonance might reveal what actually happens in the air space of the Musikvereinssaal, which, more than the other halls on our list, approaches the shape of an organ pipe.

Exercises

1. Change the location and position of the loudspeaker of your radio or phonograph. Observe the varying acoustical results in the same room.

2. Hear the variation in sound of your loudspeaker in different rooms.

3. Turn up the volume of your loudspeaker and listen to it in an adjacent room through doors alternately opened and closed. Observe how the walls and doors filter the noise.

4. Certain narrow, tall, and hard-walled (e.g., tiled) rooms, such as shower stalls, are resonant to a definite pitch. In such a room, sing a scale over the whole range of your voice until you have found the note to which the room actually resounds. (Because this experiment involves rather long waves, i.e., low tones, only men can successfully perform it.)

5. Draw the ground plan and elevation plan of halls of different shapes and sizes. Preferably choose halls in which you have a chance to hear some music. Include some unfavorable shapes, such as domed circles and octagons. Get an idea of the reflection pattern by applying geometrical acoustics, that is, by tracing the angles at which a wave bounces against the surfaces. In the case of large halls, compute the difference in the time a direct ray, and the time a reflected ray, needs to reach the ear of a listener. Place this listener in various locations in the hall.

14 | Music Theory

MAJOR AND MINOR MODES

The essence of the major and minor modes constitutes a basic problem of music, which has challenged the imagination and thoughts of music theorists through the ages. On the whole, one distinguishes two camps, both of which can boast of an impressive array of champions. One group believes that the major and minor modes are equivalent. They are reciprocal phenomena, physically produced by the reciprocal operations of division and multiplication of a string. Hence they are the musical manifestation of polarity, one of the great principles fashioning not only the outer world of nature but also the inner world of thought and imagination.

The other group believes that the minor mode is subordinate to the major mode. The major triad rules supreme, because it alone is physically produced by the overtone series of any vibrating string. The minor triad is explained as a modified, inferior, "turbid" form of the major triad.

The problem remains difficult, for apparently a complete and logically consistent system of harmony can be built on the basis of either theory. The tradition of the polarity theorists is longer, continuous, and more convincing. Its foundation goes back to Plato. The modern implications have been spelled out by such forceful advocates as Zarlino, Goethe, Riemann, Kauder, and Kayser. The turbidity theorists, on the other hand, gained popularity in the nineteenth century and therefore loom larger to the heirs in the twentieth. Helmholtz and Hindemith have been widely read.

The argument for the supremacy of the major mode proceeds from the natural phenomenon of the overtone series. Here the major triad occurs in a prominent, strong position. It is actually audible whenever a single tone is generated. The minor triad must be secondary; for in relation to the funda-

189

mental tone, it never appears in the overtone series. The minor third (in the words of Zelter) is therefore "no immediate gift of nature," and one is obliged to treat it as a "lowered major third." This clouding of the natural major triad is the reason for the melancholy effect often attributed to the minor triad. Hence the relief provided by the *tierce de Picardie*—the major third in the final chord of a composition in the "inferior" minor mode.

The following objections can be quickly raised against the argument that the minor triad is only a major triad made turbid:

(1) If the consonance of the major triad with the overtones is the main criterion, why is the minor triad not treated as a dissonance? The minor third does not occur in the overtone series against the fundamental and actually clashes with such prominent partial tones as 5, 9, 10, 15, et cetera. Yet it is generally treated as a consonance, entering without preparation and requiring no resolution.

(2) The very concept of tampering with an interval to create another one seems arbitrary. Goethe's formulation is lucid: "If the third is an interval provided by nature, how can it be flatted without being destroyed? How much or how little may one flat or sharp it in order that it may no more be a major third, and yet still be a third? And when does it cease being a third altogether?

(3) If the physical phenomenon of the overtone series is to rule supreme, why does the explanation of the major triad stop at the 6th partial? The drop in audibility between the 6th and the 7th partials is not such as to justify a sharp break at this point and a radical elimination of all higher overtones.

The argument for the equivalence of the major and minor modes proceeds from the spiritual principle of polarity. It appeals to the psychic reality of two opposites which we can hear and to which we react. Acceptance of the polarity theory does away with the first two objections raised above against the turbidity theory. Major and minor appear as equally consonant, though psychologically reciprocal; and the dignity of the minor third as an individual entity is preserved. The answer to the third objection points up the difference between a physical and a musical approach to acoustics. No satisfactory scientific answer has been found to the question concerning the unity of the senarius or the break after the 6th partial tone. In all probability, it will never be found; for by definition, scientific investigations do not eventuate in value judgments. The musician, on the other hand, not only commits himself as a humanist to make value judgments but he takes them just as seriously as scientific statements about quantities. The polarity theorist hears the reality of the peculiar musical value of the triad; he understands its scientific representation by the senarius; and he concludes that *therefore* the senarius acquires a peculiar general value.

The risk of subjectivity in any evaluation is obvious, but so must be for the musician the necessity of taking this kind of risk. Actually, a check on the objective validity of a qualitative judgment may not be entirely impossible. The contemporary Swiss philosopher Hans Kayser has devoted much of his work to the discovery of harmonical norms in various fields of science. We need only recall the examples given earlier for the formative power of the senarius to support the belief in a certain objectivity of the senaric values (cf. pp. 30 f.). Kayser has rightly criticized the prevalent scientific attitude which favors the investigation of the generative, proliferating elements of any form while accepting as given facts the limiting, shaping elements. What gives form to anything if not the interplay of a generative force that acts from within and a limiting force that acts from without? A dialectic of these two principles can supply "objective" answers to "subjective" commitments. The formative value of the senarius pervades the universe, and not just the triad.

Although the polarity theory seems to offer a more satisfactory explanation of the major-minor problem than the turbidity theory, at least two difficulties remain.

First, undertones do not exist as a spontaneous phenomenon of physical nature. This objection is irrelevant to a polarity theorist, who develops a harmonic theory not from physical phenomena but from spiritual principles. The results of number operations, such as division and multiplication, applied to the string are independent of the existence or nonexistence of parallel natural phenomena. Goethe, in a letter to Zelter of 22 June 1808, expresses his amusement at the observation that nature produces only one half of the acoustical polarity: "It is asking too much of an experiment to perform everything. . . . One should devise an experiment which could demonstrate the minor mode as being equally original." The experiment requested by Goethe consists in the application, as we have shown, of the reciprocal series of integers—acoustically not a phenomenon of physics but rather the projection of a principle.

The second difficulty is more serious. It concerns our inability to hear a chord from above. The minor triad, generated downward from C, will not be heard by us as C minor but as F minor. Riemann tried in vain to change the terminology when naming a minor triad by the generator (C) rather than by the bass root (F). We offer a hypothesis which might explain this contradiction without violating the postulates of musicianship. The hypothesis is based on a given condition into which we are born—that of tellurian gravity. Gravity, whatever its physical explanation, permeates our whole being—not only our body but certainly our total imagination. Ideally, as in the Pythagorean table, the major and minor triads spring from the same generator as a pair of identical chords in opposite directions. This absolute conception suffers as soon as the concept of high and low pitches, of altitude, in short, of gravity, enters the system. The influence of gravity does not affect the major triad, for the generator C is also the fundamental of the chord. But in the other member of the pair, in the minor triad, the generator and the fundamental become divorced. "Absolute Conception" and "Tellurian Adaptation" are in contradiction to one another. "Absolutely" we ought to hear the minor chord generated by C as C minor, but "tellurically" we do hear it as F minor. This inner schism between structure and apperception is based on polarity. This situation in music is not much different from the geotropism of plants.[1] Although the plant grows in opposite directions—the stem upward, the root downward —the flow of the sap is unidirectional, that is, "tellurically adapted."

Any hypothesis is valid to the extent to which it can supply explanations without inherent contradiction. Let us test these theoretic considerations on some examples from music literature.

What accounts for the iridescent fluctuation of major-minor in many works of Romantic composers? Consider the last of the *Lieder eines fahrenden Gesellen* by Gustav Mahler. After an opening section in the minor mode, the middle section (meas. 17 ff.) clearly begins in C major only to repeat the motif immediately in C minor. At this point, the listener is uncertain whether the ma-

[1] Hans Kayser, *Harmonia Plantarum* (Basel, 1943).

jor mode brought a momentary brightening or whether the minor mode is now understood as a clouding. The immediate continuation in C major seems to favor the latter alternative, but only for a time; for the whole section ends again in C minor (meas. 35 f.) so that the ambivalent effect remains unresolved. In the closing section of the song now initiated, the major-minor interplay begins anew. The final measures exhibit a strong cadence toward F minor so that in retrospect one thinks of the major mode as a temporary clearing-up. On the other hand, the sheer duration of the middle section in the major mode tends to weaken this feeling so that one might also hear the end as a final shrouding. Similar examples of iridescence can be found in abundance in the works of Schubert. He often writes in such a way that the major triad is heard, not as a given norm, but distinctly as a brightening of the minor triad. A polarity theorist would say that such cases present a decidedly tellurian, earthbound, use of the triads.

What accounts for the eminent role of the subdominant in literature? Here the overtones cannot help at all, for F is not present in the overtone series of C. The turbidity theorists have helped themselves by transposing a secondary interval $(G - C)$ to the tonic. This is actually an illicit operation; for with the same right one could then validate the minor third E-flat by transposing the secondary interval $E - G$ to the tonic. This inconsistency speaks strongly against the turbidity theory. No such difficulty exists in the polarity theory, where the subdominant directly emerges as having equal force with, though opposite character to, the dominant.

The harmonic behavior of chords is well explained by the recognition that two triads spring from one generator—a major triad upward, and a minor triad downward. This unfolding of a tone in both directions forms a stable whole:

Figure 66

A unilateral realization disrupts the balance and makes the chord tend toward its complement. C major tends toward F minor, and F minor toward C major, like one sex toward its complementation in the other. Hence, in general terms, the basic rule of harmony which states that each major triad tends to behave like a dominant; each minor triad, like a subdominant. This is not a mechanical rule but a real insight into the relation between tone and psyche. Musical norms are not conventional fictions but psycho-physical facts.

The oscillation thus produced is, theoretically, perpetual. The original trends remain in force and prevent us from finding an end, that is, from hearing a closing cadence:

Figure 67

A composer, whose legitimate concern is with beginning and ending, can effect a stoppage of the perpetual motion by introducing, as it were, a sudden countercurrent—in musical terms, by changing the mode of the chord meant to sound final:

Figure 68

The cadence becomes still more definite when this "counterblow" is combined with an oscillation to the opposite side of the tonic:

Figure 69

The classic cadence shows its origin when in this oscillation to the opposite side the middle term is omitted:

Figure 70

Other rules of harmony, that is, psychic realities, emerge from this fundamental recognition. Having equated the major with the dominant realms on
one side, and the minor with the subdominant realms on the other, we understand the intrusion of the major dominant into the minor mode, and that of
the minor subdominant into the major mode. The former is so common that
the illustrations may all be drawn from the reader's own experience. The
mere term "harmonic minor" vouches for the general acceptance of this idea.

The role of the minor subdominant in both modes, equally well manifested
in music literature, has been somewhat slighted in traditional theory books,
probably for tellurian reasons. In the minor mode, of course, the subdominant
acts with a force comparable in the major mode to the major dominant. Many
Brahms compositions end this way (for example, Rhapsodie, op. 119, no. 4).
The minor third in the subdominant chord resolves downward like a leading
tone, exactly reciprocal to the resolution upward of the major third in the dominant chord. In the major mode, the minor subdominant produces cadences
like the following at the end of the second movement of the Third Symphony
by Brahms:

Figure 71

Here, the major subdominant in the first half of the cadence is strengthened in
the second half by appearing as the relative of the minor subdominant (A-flat
major for F minor). One can go so far as to say that the crucial tone A-flat
here forms its own triad.

The most obvious use of the minor subdominant in the major mode is demonstrated by the diminished seventh chord. Tellurically, this chord is normally

explained as a dominant ninth without root. Why flatten the ninth unless it be a representative of the minor subdominant?

Figure 72

The characteristic force of this chord is derived from the united presence of major-dominant and minor-subdominant elements:

Figure 73

This remains true whether one explains the chord tellurically or reciprocally. In either case, it is a ninth chord without root, but only the polarity theory accounts satisfactorily for the A-flat instead of the A-natural.

Figure 74

The authentic resolution of the chord shows the exactly reciprocal behavior of its components:

Figure 75

One more example to emphasize the point that the behavior of chords is not a mere textbook convention but the expression of an acoustical truth. The intensity of the dominant chord and of its drive to the tonic is significantly increased, as we know, by the addition of a seventh, a note which preserves the harmonic unity of the chord while yet introducing an element easily associated with the subdominant function (*F* in C major). The reciprocal chord is:

Figure 76

In tellurian adaptation, we recognize it as the minor subdominant with an added sixth, Rameau's famous *sixte ajoutée*. The upward resolution of the sixth, corresponding to the downward resolution of the seventh, is a clue to the origin of this crucial note:

Figure 77

The complete cadence including the minor subdominant with the added sixth and the major dominant with the added seventh takes the following shape in the *German Requiem* by Brahms (No. 4, meas. 149-153):

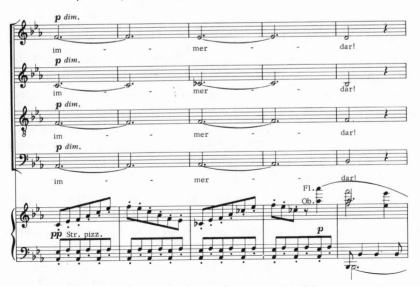

Figure 78

Further probing belongs in a theory of harmony, which lies outside the scope of this book. Two facts should be evident by now: rules of harmony, properly understood, reflect universal truths; they make sense if heard, not as conventions, but as values.

Exercises

1. In the songs by Schubert, find examples of fluctuations between the major and minor modes. Explain each occurrence in terms of polarity or tellurian adaptation, as the case may be.
2. In the literature, find examples of major-mode cadences that employ the minor subdominant chord with the added sixth.
3. In compositions of different periods, find the progression s_7-D^7 (minor subdominant with added sixth to major dominant with added seventh).
4. The deceptive cadence (V-VI) in major substitutes the tonic relative for the expected tonic. According to the turbidity theory, the same progression in minor reaches the subdominant relative—an apparently strange substitution. Explain how according to the polarity theory the deceptive cadences in both major and minor proceed in an analogous manner.

CONSONANCE AND DISSONANCE

The words *consonare* and *dissonare*, literally translated, mean, respectively, 'to sound together' and 'to sound apart.' A "sounding together" implies a close identification; a "sounding apart," a pull toward a new sound. In one case, the participating tones are at rest with each other; in the other, their dissimilarity sets up a drive toward resolution and entelechy. "The Similar and the Like did not need Harmony, but the Dissimilar and the Unlike had to be necessarily united by Harmony in order to endure in the Cosmos," we read in Philolaos, the contemporary of Socrates.[1]

In musical terms, we can identify consonance with a perfect balance of forces, with a condition of rest, with the potential of finality. Analogously, we identify dissonance with a struggle of energies, with unresolved tension, with the promise of continuation.

---

[1] Here quoted after Hans Kayser, *Abhandlungen zur Ektypik harmonikaler Wertformen* (Zürich and Leipzig, 1938), p. 82. Transl. by the authors.

In this sense, there is only one perfect consonance between two tones: the unison; and there is only one perfect consonance among chords: the triad. All other intervals and chords are more or less dissonant, varying according to the energy with which the participating forces "pull apart." The underlying assumption is some musical unity, a oneness of sound, in which the partaking tones lose their identity. Acoustics offers us two prominent manifestations of such a unity: the oneness of the string which vibrates simultaneously as a whole and in parts, and the oneness of the harmonic series which forms one tone. As this consonance is "sounded apart," it becomes ever more articulated and differentiated until the increasing separate individuation of the partaking tones makes us speak of "dissonance." The prevailing unity at any time can be heard and understood as an extension of the unison or, as the case may be, of the triad.

The whole inquiry into the nature of consonance and dissonance lies at the border between acoustics and music theory. All efforts in this area have been motivated by the understandable desire to reconcile the psychological experiences and the physical facts. The mere multitude of existing theories, from the earliest times to the most modern speculations, points up the discomfiting fact that no one theory has been generally accepted. Some questions always seem to escape the network of an explanation, regardless of one's approach. Our own suggestions and arguments will fare no better.

Our first task is to contemplate a hierarchy of intervals in regard to their inherent degree of consonance and dissonance. The qualitative difference among intervals, which we can hear, corresponds to a measurable difference, which we can calculate. We may proceed psychologically, from the division of the string; or physically, from the overtone series. In either case, a definite rank order emerges.

Psychologically, the norms are given by the order in which the various intervals appear as a result of the divisions by successive integers. We are using the word "psychologically" deliberately in order to emphasize again the correspondence of laws outside and within us, of norms in nature and in man, of forces in the cosmos and in the psyche. The hierarchy of intervals reflects not just a mathematical organization but a structure of our soul. The hierarchy can be ascertained quite methodically. Let us write down the successive fractions and identify each by the corresponding tone value as measured from $1/1 = c$. The senaric values (i.e., the ratios based on the first six numbers and their products) serve as a normative guide. Following them up to 16, and without repeating identical pitches, we arrive at a definite order. The following table, for visual intelligibility, arranges on a square grid of 16 all these fractions that issue a new interval. The table is to be read as a continuum from left to right:

$\frac{1}{1} \cdot c$

$\frac{1}{2} \cdot c^1$

$\frac{1}{3} \cdot g^1 \quad \frac{2}{3} \cdot g$

$\frac{1}{4} \cdot c^2 \qquad \frac{3}{4} \cdot f$

$\frac{1}{5} \cdot e^2 \quad \frac{2}{5} \cdot e^1 \quad \frac{3}{5} \cdot a \quad \frac{4}{5} \cdot e$

$\frac{1}{6} \cdot g^2 \qquad\qquad \frac{5}{6} \cdot eb$

$\frac{1}{8} \cdot c^3 \qquad \frac{3}{8} \cdot f^1 \qquad \frac{5}{8} \cdot ab$

$\frac{1}{9} \cdot d^3 \quad \frac{2}{9} \cdot d^2 \qquad \frac{4}{9} \cdot d^1 \quad \frac{5}{9} \cdot bb \qquad\qquad \frac{8}{9} \cdot d$

$\frac{1}{10} \cdot e^3 \qquad \frac{3}{10} \cdot a^1 \qquad\qquad\qquad \frac{9}{10} \cdot d$

$\frac{1}{12} \cdot g^3 \qquad\qquad \frac{5}{12} \cdot eb^1$

$\frac{1}{15} \cdot b^3 \quad \frac{2}{15} \cdot b^2 \qquad \frac{4}{15} \cdot b^1 \qquad\qquad\qquad \frac{8}{15} \cdot b$

$\frac{1}{16} \cdot c^4 \qquad \frac{3}{16} \cdot f^2 \qquad \frac{5}{16} \cdot ab^1 \qquad\qquad \frac{9}{16} \cdot bb \qquad\qquad\qquad\qquad \frac{15}{16} \cdot db$

Figure 79

Reducing all occurring intervals to the span of one octave, we may read off (always from left to right, i.e., in a continuous line) the following rank order of tones in relation to $c = 1/1$:

$$\frac{1}{1} \qquad \frac{1}{2} \qquad \frac{2}{3} \qquad \frac{3}{4} \qquad \frac{4}{5} \qquad \frac{3}{5} \qquad \frac{5}{6} \qquad \frac{5}{8} \qquad \frac{8}{9} \qquad \frac{5}{9} \qquad \frac{8}{15} \qquad \frac{15}{16}$$

Figure 80

Translated into general terms, the following hierarchy of intervals emerges:

> Unison
> Octave
> Fifth
> Fourth
> Major third
> Major sixth
> Minor third
> Minor sixth
> Major second
> Minor seventh
> Major seventh
> Minor second

From the one perfectly consonant interval of the unison, the intervals become gradually more dissonant. There is no sharp dividing line between consonance and dissonance. The fifth, for instance, is more dissonant than the octave but less so than the fourth. We can state the same fact by calling the fifth less consonant than the octave but more so than the fourth. The grouping of consonances and dissonances in two distinct camps, as practiced by our common jargon, is at best a convention which has changed over the centuries. Today the line has been placed just above the seconds and sevenths; in the thirteenth century it was placed above the thirds and sixths. Any line drawn anywhere in this hierarchy is arbitrary. The increase in tension is gradual. The "pulling apart," the dissonance, becomes steadily stronger. Only the unison is at perfect rest. No arbitrary convention can ever change this fundamental law.

The critical student will make at least three observations when studying the table of intervals (Fig. 79) that he has just developed.

First, one tone in our language has not been accounted for. The tritone *c-f♯* is missing. It lies so "far out" that a division of the string by 45 is necessary to have. it materialize; and even then the proportion is complex: 32/45, as inferrible from 4/5 × 8/9, major third plus major second. Both 32 and 45 are senaric products, which means that the tritone definitely belongs to our musical vocabulary. But the inherent tension of the tritone is considerably greater than that of the nearest dissonance above it, the minor second. The tritone appears as the strongest possible dissonance in our twelve-tone system, remote from the rest, a real *diabolus in musica*.

Second, two tones turn up with conflicting numerical values. In one instance, *d* is 8/9; in another, 9/10. Also, *b*-flat materializes at 5/9 and at 9/16. Does the existence of two different measures for the same tone threaten the identity and precision of the pitch? Obviously two different, although closely adjacent, spots on the monochord string are involved. In either case with octave reduction, $d = 8/9$ is reached by two fifths ($2/3 \times 2/3 \times 2/1$); $d = 9/10$, by a fourth and a sixth ($3/4 \times 3/5 \times 2/1$). In either case, one calls the result *d*, although this *d* evidently has two distinct personalities. The situation is similar in regard to *b*-flat, which is split into 5/9 (a major third below two fifths) and 9/16 (two fourths).

How far apart do the two *d*'s lie? 8/9 ÷ 9/10 = 80/81. What is the discrepancy between the two *b*-flats? 5/9 ÷ 9:16, or again 80/81. This small measure is called the "syntonic comma." Minimal when compared to the smallest step in our scale, the half-tone 15/16, it is yet clearly audible on the monochord. On a string of 120 centimeters, for instance, the two *d*'s lie 1.67 centimeters apart. Sing *d* in the dominant chord of a C major cadence and hold it while harmonizing it by the two subdominant chords (IV with the added sixth, and II)— you will experience the pull.

One cannot forgo speculation on the interesting fact that precisely at the whole tone we are suddenly given two conflicting norms of equally pure lineage, and that we cannot *remember* usefully their distinctness. Clearly, we hear the difference between 8/9 and 9/10 when a good violinist, for instance, or a singer makes the smaller whole tone succeed the larger at the beginning of an ascending major scale. Out of context, however, we are content to think of either interval as a whole tone. We can hear the variable relation, but we cannot remember the aural absolute. Possibly the line drawn by common verbal usage between the "perfect and imperfect consonances" and the "dissonances" derives a functional meaning beyond its apparent arbitrariness from the first appearance of a variable interval precisely at the whole tone.

Third, although, in relation to *c*, *e* arises before *a*, a discrepancy exists between the major tenth and the major third. As is clear from the table (Fig. 79), the major sixth falls between the two. A major third may thus be called more consonant than the major sixth only if we do not practice octave reduction. In precise terms, the major third (but not the major tenth) is less consonant than the major sixth. One becomes sensitized to this distinction when writing traditional harmony exercises in which one rightly endeavors to avoid the placement of the major third directly above the bass. The point has been made persuasively that the inherent tension of a tight major third and its eventual expansion to a more consonant major tenth (and octave of the tenth) form the musical contents of the entire first prelude of the *Well-tempered Keyboard*.[2] A similar ambivalence exists, in regard to the minor seventh, between the major ninth and the major second. Without impairing the truth of the earlier diagram, which adhered to the principle of octave reduction, the following modification is justified for the sake of greater precision:

> Unison (1/1)
> Octave (1/2)
> Twelfth (1/3)
> Fifth (2/3)
> Fourth (3/4)
> Major tenth (2/5)
> Major sixth (3/5)
> Major third (4/5)

2 Heinrich Schenker, *Fünf Urlinie-Tafeln* (New York, 1932). Compare the distribution of the notes in the first and last chords.

Minor third (5/6)
Minor sixth (5/8)
Major ninth (4/9)
Minor seventh (5/9)
Major second (8/9)
Major second (9/10)
Minor tenth (5/12)
Major seventh (8/15)
Minor second (15/16)
Tritone (32/45)

Thus far we have proceeded from the division of the string. If we now take the overtone series as a starting point, the hierarchy of intervals emerges from the order in which each successive interval makes its appearance. The underlying oneness, which we initially stipulated as a condition of consonance, is here provided by the very nature of the phenomenon. Tone, while a complexity, is perceived as unity as long as the overtones remain in their ordinal position in regard to both distance and loudness. This oneness is destroyed when an overtone changes its position toward the fundamental through octave reduction and a concomitant increase in loudness.

As we ascend the overtone series, there is no question about unison, octave, fifth, and fourth. At the 5th partial, two new intervals spring up: the major third (or tenth, if you will) against the fundamental, and the major sixth against the 3rd partial. The suggestion is reasonable that the priority of intervals be established by the relation of each newly materialized overtone to the preceding partials from the bottom upward. The 5th partial forms first the tenth (actually the octave of the tenth), then the major sixth, and finally the major third. Hence we are correct in stating that the major sixth is more consonant than the major third, but less so than the major tenth. This is the same kind of ambiguity that we encountered at the comparable spot in our earlier, psychological, approach to the same problem of hierarchy. Octave reduction is practical but it sometimes invades our thinking at the expense of precise definitions.

Proceeding along these lines up the overtone series, while excluding those partials that lie outside our tone system, we can establish the following rank order among intervals from the perfect consonance of the unison to ever increasing dissonances:

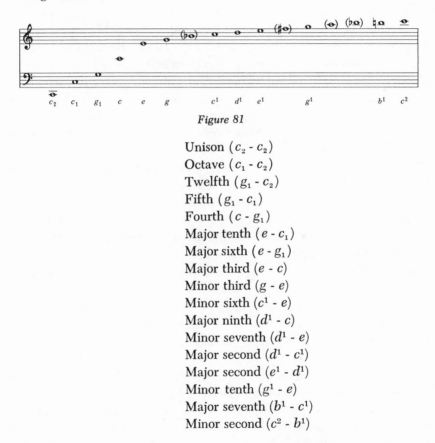

Figure 81

Unison $(c_2 - c_2)$
Octave $(c_1 - c_2)$
Twelfth $(g_1 - c_2)$
Fifth $(g_1 - c_1)$
Fourth $(c - g_1)$
Major tenth $(e - c_1)$
Major sixth $(e - g_1)$
Major third $(e - c)$
Minor third $(g - e)$
Minor sixth $(c^1 - e)$
Major ninth $(d^1 - c)$
Minor seventh $(d^1 - e)$
Major second $(d^1 - c^1)$
Major second $(e^1 - d^1)$
Minor tenth $(g^1 - e)$
Major seventh $(b^1 - c^1)$
Minor second $(c^2 - b^1)$

The tritone, here as before, is absent. The devil in music as elsewhere seems to have been cast out from the established order, emitting his threatening sound from the bottom of the hierarchy. In this and all details, the total result is identical with the one reached earlier. In the first case, an intellectually controlled experiment with man in the center; in this case, the naturally spontaneous phenomenon of overtones—the same norms govern the psychic and the physical realms. This must be so, or man would forever be at a loss to relate himself to the world, to understand the wholeness of his inner life within the outside surroundings.

The identity of the results gives them the appearance of being true and conclusive. Perhaps they are. Various questions, however, remain worrisome unless one concedes—in line with the general twentieth-century attitude toward consonance and dissonance—that "a psycho-structural event of such radical nature as harmony . . . cannot be fully explained by any single principle but rather by several, quite differently disposed principles geared to one another."[3]

Here is a sample list of possible objections to the hierarchic structure outlined above:

Why does a fifth that is slightly out of tune still sound consonant? On a well-tempered keyboard, for example, the ratio of the fifth (293/439) is certainly "worse" than that of even the tritone (32/45); yet we hear it as a consonance.

Why does the Pythagorean third (8/9 $\times$ 8/9) have a more complex ratio (64/81) than the major second (8/9), the major sixth (16/27), and the tritone (32/45), notwithstanding a higher degree of consonance?

Does c-f♯, played on the piano, for instance, really sound more dissonant than c-d♭ or c-b?

Is c-g♯ more consonant when heard as two superimposed major thirds (16/25) rather than as an alteration of a perfect fifth?

Does a bare major seventh (c-b) become more consonant when joined by an intermediate g (c-g-b)?

The first two of these questions defy any explanation by theories based on ordinal numbers and overtones without actually invalidating them. We hear, not physiologically, but psychologically. The idea of a fifth or of a major third is stronger than the materialization, which even under the best circumstances remains imperfect. We hear the norm and relate the physical sensation back to it. Beyond a certain threshold, of course, a flattened fifth reaches a point where we no longer call it "tempered" but "out of tune," i.e., "dissonant." The

[3] Albert Wellek, "Die Aufspaltung der 'Tonhöhe'," *Zeitschrift für Musikwissenschaft*, XVI/11-12 (November-December, 1934), 541. Authors' translation.

deviation of the Pythagorean third from the norm is so small as to remain inside the critical threshold.

The next two questions point up an age-old dilemma that cannot be resolved below the level of faith. The conflict crystallized in Classic Greece between the schools of Pythagoras and of Aristoxenos, Aristotle's pupil. Against the tradition of the Pythagoreans, the Aristoxenians postulated that in music the ear is the sole arbiter. Whose ear? one readily inquires. If a tritone sounds less dissonant than a major seventh to one person, how can he convey this subjective reaction to another person for whom it need not have any validity? The ultimate hope rests in a theory that reconciles the distinction between the inside ear and the outside world by recognizing that both are governed by the same laws.

The last question, finally, while Aristoxenian in concern, leads away from intervals to a consideration of the consonance-dissonance problem in chords. The essential insight into rank order to which the investigation of intervals has given us access is here more difficult to isolate though equally significant. The entirety of an interval stems from two formal concepts: relationship and wholeness. The interval stretches between the two limits set by the tones which alone render it real. In the moment of sounding, however, these limits vanish, and we hear the wholeness of the musical interval. A triad, similarly, can be described in terms of relations among the fixed points. Yet, the musician listening to the sound of the triad will accept it in its entirety rather than as a summation of intervals.

Hence the triad as given by the senarius and the harmonic series vies with plain intervals for primacy in determining the consonant or dissonant value of a musical event. A chord gains in consonance the more closely it adheres to the natural position in the overtone series in regard to both distance and loudness. A deviation from the unity of the natural model gives individuality to a tone. The deviation may vary from a relatively simple octave reduction to a definitive departure. An individualized tone "sticks out" and thereby destroys the oneness of the musical event. This is the reason for the acknowledged practice of the masters—and for the lesson to us—not to crowd the triad tones into a low octave range, or not to double the major third in the triad at the expense of root and fifth. In either case, the individuation impairs the consonance—in the first instance by a shift of position, in the second by a protrusion of dynamics. An inversion, for similar reasons, indicates a gain in dissonance.

Among the many theories, there is hardly any disagreement concerning the inherent consonance of the triad. The senarius and the overtone series both point to the same conclusion. The turbidity theorist will make reservations about including the minor triad in this statement; but the conclusion is thereby not contradicted, for to the turbidity theorist, the minor triad is not an autonomous triad, anyway.

Without denying the impression of restful finality produced by the triad, sensitive musicians have rightly wondered about situations in which the triad is used to produce an impression of tension; there is no resting on the B-flat major triad in the following phrase:

Figure 82

They may also rightly wonder about situations in which chords that are not perfect triads are yet being successfully used as closing chords:

Figure 83

Because these observations deal with typical experiences (the examples can be easily multiplied) rather than with exceptional, they lead us to the following summary:

The triad is consonant.

All other chords are dissonant.

The triad may be used as a dissonance.

Other chords—maybe all of them—may be used as a consonance.

This juxtaposition seems to point to the existence of two different sorts of consonance quality and of dissonance quality. One sort is natural and inherent in the phenomenon; our attitude toward it is therefore rather passive. The other sort is psychological and imagined by us as a current inducted into the phenomenon; our attitude toward it is therefore extremely active. Because the jumbling of both sorts has been the source of unfortunate errors, the separation by an appropriate terminology is most advisable. We suggest that the term-pair "consonant-dissonant" be reserved for the natural qualities, and that the term-pair "static-dynamic" be adopted for the parallel interpretive qualities (cf. "ontic-gignetic," pp. xi-xii).

At this place, we do well to quote some thoughts by Kayser about "the concept of the 'dynamic' which plays a decisive role in the modern world. In physics, matter has been dissolved to become a dynamic concept of energy; the originally 'statically' perceived atom has been dissolved into a spatial configuration of electro-dynamic units of energy which can no longer be grasped by apperception. All technology is a single field of applied dynamics. In biology, materialistic dynamism is transformed into 'vitalism' : life is movement and carries its justification in itself, i.e., in the unrest of the one who moves. In philosophy, dynamism becomes voluntarism. Here the tendencies toward the voluntarizing of all acts of knowledge have become so strong that to occupy one's self with 'eternal ideas' in Plato's sense seems almost disreputable. In the arts, particularly in modern music, dynamics has overgrown everything to such an extent that one wonders why this 'motorization' still employs the usual music instruments instead of replacing them with 'timely' sound producers such as sirens, car horns, gasoline motors, hand grenades, noise makers, and other explosives.[4] Even the concept of God as a 'living' creator, ever 'renewing' himself in the world and in the soul of man, has been woven more and more into a dynamic action and thereby placed in jeopardy of becoming a merely vitalizing principle. . . .

"The contribution of harmonics to a rectification, i.e., to a natural directing and ordering, of dynamics consists, generally expressed, in the coordination of the static form of the tone value to the dynamic vibration of the tone number.

[4] Less than two decades after Kayser wrote these words, one no longer wonders. Events have caught up with the gruesome prophesy.

Because this coordination exists *a priori* in the archetypal phenomenon of tone, where a value is born with each frequency, the danger of the dynamic frequency running amuck is inhibited from the onset, tamed by 'measure' or, in other words, lifted into the sphere of morphology."[5]

The range of a possible static-dynamic interpretation of tone experiences, while being wide, yet probably is not boundless. Apart from any theoretical considerations, recent developments in the aesthetics and techniques of musical composition seem to suggest a limit beyond which the norms cease to be recognizable. In the period after the First World War, when the existence of consonance and dissonance was largely disregarded or even denied, the progressive psychologization of music relied solely on the concept pair "static-dynamic" for producing the desired effects of "binding and unbinding." Logically enough, in that period the triad was ostracized. In more recent years, however, the feeling has begun to prevail that composers had gone too far; for the artificially constructed norms threaten to drive music into becoming a secret language evolved from, and intelligible only to the holders of, a particular code system. The best hope for today and for the immediate future lies with those composers who incorporate the rich discoveries made during the years of frantic experimenting into styles of writing based on inherent natural norms. A mere feeling of revolt against natural principles is inadequate to create a good work of art. Norms are frames of reference without which the concept of artistic freedom becomes meaningless and which alone ensure the right kind of freedom.

Exercises

1. The following exercises concern only the purely acoustical, natural aspect of the problem, and not the musical, psychological aspect.

a. Remembering that a fourth is less consonant than a fifth, play on a keyboard instrument first ♯ 𝄞 𝅝 and then ♯ 𝄞 𝅝 The dissonant implications of the fourth disappear, because the two tones are now heard and understood as functions of the fundamental.

[5] Hans Kayser, *Lehrbuch der Harmonik* (Zürich, 1950), pp. 9 f. Transl. by the authors.

b. Play [score] and then [score]

The effect of the minor third disappears.

c. Play [score] and then [score]

The effect of the major sixth disappears.

2. Big spaces between tones decrease the inherent degree of dissonance, because the relative position of the tones thereby approaches the natural position of dissonant partials.

a. Play in succession and compare:

b. Play in succession and compare: [score]

c. Play in succession and compare: [score]

3. On the monochord, find the two different values for d and the two different values for b-flat, and hear the syntonic comma.

4. Is the distance on the monochord string the same between the two d's as between the two b-flats?

5. Sing the tone d while harmonizing it on the piano in different ways within an overall C major tonality.

6. Find examples in the literature where the major triad is used as a dissonance.

7. Find examples in the literature where chords other than the triad are used as consonances.

TEMPERAMENT

Temperament is an attempt to reconcile two conflicting musical norms. One is the octave, which sets a definite, frame-like limit. The other is the bypassing of the octave by the accumulation of any other primary, scale-building interval. Temperament is the musical way of fitting an otherwise endless series into a definite space.

The octave is a microcosm, because any of the infinite number of possible tones can be projected into a single octave space without losing its identity. As soon as one accepts the octave—that "basic miracle of music," as it has been called—the task presents itself of filling it with a discrete number of tones. This task could conceivably be approached by one of two ways. We might divide the octave space by an arbitrary number of steps, or we might select some harmonic ratio and let the corresponding interval generate new tones by a series of superpositions. The first method can be called melodic; the second, harmonic. Whatever the approach, melody and harmony are so intimately bound up with each other that finally one will be inevitably adjusted to the other. We submit that theory favors, and Western history confirms, the primacy of harmonic norms. The octave itself is a harmonic, not a melodic, discovery. The same process by which the octave has been carved from infinity can be logically continued, eventuating in finite steps within the micro-infinity of the octave. We shall therefore have to select some primary harmonic entity as a building stone for the production of a definite number of tones within the octave. The result will be a scale.

In the Pythagorean system, that primary entity is the interval next to the octave in the harmonic hierarchy, namely, the fifth. Perhaps the first scale (or at least the first fixed tones of a scale) resulted from the process of flanking a generator by its upper and lower fifths, its dominant and subdominant:

Figure 93

We have referred to "fixed tones," because the interplay between the harmonic norm and the dynamism of melody has at all times produced secondary, "movable" pitches beside the fixed ones. Thus, for instance, the fixed tones in Greek music were represented by the span of the tetrachord; the movable tones, by the diatonic, chromatic, or enharmonic steps that filled that span. In our own system, the fixed tones are represented by the diatonic scale; the movable tones, by the chromatic alterations.

The addition of another fifth above and below the generator produces the pentatonic scale:

Figure 94

One further expansion by reciprocal fifths creates the Pythagorean scale of seven tones:

Figure 95

One notes that this symmetrical arrangement of fifths results in the Dorian mode, itself perfectly symmetrical. The two tetrachords (c - f, and g - c^1) are identical, the semitone lying in the center of each. The ratios of this scale, in relation to the fundamental, are the following:

1/1	8/9	27/32	3/4	2/3	16/27	9/16	1/2
c	d	e♭	f	g	a	b♭	c¹

The ratios of the Pythagorean intervals are of interest. We obtain them by systematically comparing the differences between a larger and a smaller interval:

The semitones

$$e♭ - d = 27/32 \div 8/9 = 243/256$$
$$b♭ - a = 9/16 \div 16/27 = 243/256$$

The whole tones

f - $e\flat$ = 3/4 ÷ 27/32 = 8/9

g - f = 2/3 ÷ 3/4 = 8/9

a - g = 16/27 ÷ 2/3 = 8/9

c^1 - $b\flat$ = 1/2 ÷ 9/16 = 8/9

The minor thirds

$e\flat$ - c = 27/32 ÷ 1/1 = 27/32

f - d = 3/4 ÷ 8/9 = 27/32

$b\flat$ - g = 9/16 ÷ 2/3 = 27/32

c^1 - a = 1/2 ÷ 16/27 = 27/32

The major thirds

g - $e\flat$ = 2/3 ÷ 27/32 = 64/81

a - f = 16/27 ÷ 3/4 = 64/81

The fifths

g - c = 2/3 ÷ 1/1 = 2/3

a - d = 16/27 ÷ 8/9 = 2/3

$b\flat$ - $e\flat$ = 9/16 ÷ 27/32 = 2/3

c^1 - f = 1/2 ÷ 3/4 = 2/3

This Pythagorean tuning, with the fifth as the measure of the microcosm within the octave, is perfect for melodic purposes. But the fifths expanding in either direction never meet the octave, for 12 fifths reach slightly farther than 7 octaves. The mathematical proof reads:

$$(2/3)^{12} \div (1/2)^7 = 2^{19}/3^{12} = 524288/531441 < 1.$$

The difference between the two pitches, as expressed by the last fraction, is called the "Pythagorean comma."

Post-Pythagorean tuning, including our own, takes its measure, not from the fifth, but from the triad. In contradistinction to the Pythagorean "ternarius," the triad introduces the senarius; for the beauty and perfection of the triad are dependent on the third, which corresponds to the number 5.

The first result to be observed is a simplification of the ratios. Many tones can be reached by a shorter way with the help of thirds and fifths than with the help of fifths alone. The major third itself, to give the most obvious example, becomes a direct relationship instead of the product of four fifths. The major seventh can be attained by a fifth plus a third (*C* - *G* - *B*, or *C* - *E* - *B*) rather than by the longer way across five fifths (*C* - *G* - *D* - *A* - *E* - *B*). The ratio of the half-tone becomes 15/16 as against 243/256.

Our system, then, arbitrarily called "just intonation," presents itself as follows:

Figure 96

It is defined by the three triads of tonic, dominant, and subdominant. The scale is the melodic projection of these three chords; it contains the different tones of the chords, and only these. Here are the ratios of the scale:

1/1	8/9	4/5	3/4	2/3	3/5	8/15	1/2
c	*d*	*e*	*f*	*g*	*a*	*b*	*c*¹

Here are the ratios of the intervals in just intonation:

The semitones
$$f \text{ - } e = 3/4 \div 4/5 = 15/16$$
$$c^1 \text{ - } b = 1/2 \div 8/15 = 15/16$$

The whole tones
$$d \text{ - } c \qquad\qquad = 8/9$$
$$e \text{ - } d = 4/5 \div 8/9 = 9/10$$
$$g \text{ - } f = 2/3 \div 3/2 = 8/9$$
$$a \text{ - } g = 3/5 \div 2/3 = 9/10$$
$$b \text{ - } a = 8/15 \div 3/5 = 8/9$$

The minor thirds
$$f \text{ - } d = 3/4 \div 8/9 = 27/32$$
$$g \text{ - } e = 2/3 \div 4/5 = 5/6$$
$$c^1 \text{ - } a = 1/2 \div 3/5 = 5/6$$
$$d^1 \text{ - } b = 4/9 \div 8/15 = 5/6$$

The major thirds

$$e \text{ - } c = 4/5 \div 1/1 = 4/5$$
$$a \text{ - } f = 3/5 \div 3/4 = 4/5$$
$$b \text{ - } g = 8/15 \div 2/3 = 4/5$$

The fifths

$$g \text{ - } c = 2/3 \div 1/1 = 2/3$$
$$a \text{ - } d = 3/5 \div 8/9 = 27/40$$
$$b \text{ - } e = 8/15 \div 4/5 = 2/3$$
$$c^1 \text{ - } f = 1/2 \div 3/4 = 2/3$$
$$d^1 \text{ - } g = 4/9 \div 2/3 = 2/3$$
$$e^1 \text{ - } a = 2/5 \div 3/5 = 2/3$$

The two semitones are identical, and the major thirds are all pure. But unlike the Pythagorean system, just intonation proposes two sizes of whole tones: $8/9$ (d - c) and $9/10$ (e - d). Of the minor thirds, one has remained Pythagorean (f - d); it is smaller than the normal minor third of our system. The fifths are pure except one, which destroys the perfection of the major and minor triads on d. Composers have sometimes shown their sensitivity to the imperfection of the fifth a - d. Anton Bruckner reportedly warned his students to treat this particular fifth as a dissonance in C major.

The difference between the two whole tones, or between the two minor thirds, or between the two fifths, is called the "syntonic comma." It equals $80/81$, as the following calculations show:

$$d \text{ - } c \; vs. \; e \text{- } d = 8/9 \div 9/10 = 80/81$$
$$g \text{ - } e \; vs. \; f \text{ - } d = 5/6 \div 27/32 = 80/81$$
$$g \text{ - } c \; vs. \; a \text{ - } d = 2/3 \div 27/40 = 80/81$$

If we said that the Pythagorean system of intonation is perfectly adapted to melodic use, we cannot claim that just intonation is equally successful for harmonic use. In fact, even the different modes are unsatisfactory in just intonation; for difficulties similar to those encountered in the major scale present themselves in the minor, Dorian, or any other scale one might build on the same fundamental. In C minor, for instance, the step from e-flat to f is smaller than that from f to g: and the minor third d to f is smaller than the other occurring minor thirds.

In fine, whereas the Pythagorean tuning does not give us the perfect triad, in just intonation some intervals and chords are badly out of tune. Yet all these shortcomings are but a foretaste of the difficulties that await us when we try to transpose the modes, that is, when we try to modulate.

One example may suffice for an illustration. Suppose we transpose the pure third 4/5 by pyramiding it onto itself:

1/1	4/5	16/25	64/125
c	*e*	*g♯*	*b♯*

Let us try the same operation with whole tones (actually with every second fifth):

1/1	8/9	64/81	512/729	4096/6561	32768/59049	262144/531441
c	*d*	*e*	*f♯*	*g♯*	*a♯*	*b♯*

The two *b*-sharps thus reached are different from each other, and neither is identical with *c*.[1] Compared to the octave (1/2, or 60 cm on the sonometer with a string length of 120 cm), the *b*-sharp reached by thirds is flat (64/125, or 61.44 cm) whereas the *b*-sharp reached by whole tones is sharp (262144/531441, or 59.19 cm).

Here we may grasp the reason for the impossibility of tuning so that triads from all tones will be pure, and for the assumption of different sizes by the same interval depending on how one reaches it. Every piano tuner operates according to the first limitation, and every singer and violinist according to the second. The musician must realize that no interval added to itself any number of times can ever regain the starting tone in any octave range. Mathematically speaking: no power of any number other than 2 can ever equal a power of 2.

Let us summarize the situation. Nature provides us with two conflicting phenomena. On the one hand, it offers us an infinity of tones and hence an infinity of intervals. On the other hand, it gives us a definite framework, the octave, capable of containing that infinity. From among the unlimited number of intervals, we select a few as harmonic-melodic norms. But a row of such norms never fits the octave. At this point we do well to pause and reflect upon this remarkable inner contradiction. We might interpret it as a manifestation of two opposing forces—one extensive, the other formative. The one creates matter; the other, shape. If matter were allowed to multiply unchecked, annihilation of nothingness would be the final result—something that in turn would be the equivalent of nothingness, a kind of universal cancer. The formative principle opposes such a process. The result is individuation, an entity suspended between nothingness and a-nothingness. We can detect the working of these two forces in the genesis of the musical work as well. In order to produce that individuality characteristic of a work of art, musical matter and musical form must balance each other. The two forces of growth and of growth limitation are another manifestation of polarity.

The general musical situation thus described calls for a compromise, consisting in a deviation from the norms so that a certain number of intervals may exactly fit the octave. This operation is called "tempering"; the resulting scale, a "tempered" scale. Accordingly, temperament should be viewed as an attempt to represent the indefinite within the definite, to reduce infinity to finite limits. If temperament can rightly be called a compromise, it may also be considered a step in the stylization of musical matter. As such it confers to the immediate raw material of music, namely, to the scale, the dignity of an art product, thereby reaffirming that music is, before and above all, an affair of the spirit.

Temperament, then, consists in a deviation from norms. A tempered keyboard, for instance, renders almost all intervals as mere approximations. Why do our ears not consider this kind of situation intolerable? The answer lies in the meaning of the term "norm." In contradistinction to "laws" (e.g., "natural laws"), norms permit approximation. We are able to recognize a norm though it may not be embodied to perfection. Were it not so, music-making would prove impossible, for exact intonation exists only theoretically. Tones, intervals, and chords as actually perceived are mere suggestions of what is meant we should hear. We hear psychologically rather than physiologically. As in geometry, we perceive the norms through more or less imperfect representations. We mentally correct the impressions received through the senses, and we do so always in the direction of some norm. We recognize a point even though none ever put by man on paper has possibly met the geometric definition of a point. In music, the norm is determined by the grammar of a style, that is, by the context. On a keyboard instrument, D-sharp and E-flat have physically the same pitch (just as in spoken language, homonyms like "bare" and "bear" have acoustically the same sound). The context relieves any doubt and tells us how to interpret what we hear. If you listen with musical sensitivity to the following chords played on a piano, you will hear g-sharp as having a higher pitch than a-flat:

Figure 97

Or, feel the pull, upward and downward, singing a sustained F while playing the following different harmonizations of the same note as done by Verdi in the last scene of *Falstaff* (between rehearsal numbers 28 and 29):

Figure 98

We may go about tempering a scale in a number of ways. It all depends on what values we wish most to preserve among the norms. This conclusion is tantamount to a declaration that the kind of temperament will vary with the great style categories of music. In some styles, no necessity for temperament arises. Pure Pythagorean tuning was perfectly adequate for the monodic music of the old Greeks without, or with few, transpositions. Just intonation, on the other hand, is inadequate for chordal music even without any transposition. The widening of the tonal movement by modulation in the sixteenth and seventeenth centuries eventually and necessarily led to the general acceptance of our equal temperament. Bach's "well-tempered" keyboard is an early manifestation of the new style, notwithstanding the traditional aspects of the work. Among the various proposals for tempered tuning before the establishment of equal temperament, only one, the so-called "meantone temperament," proved itself successful. We shall therefore consider these two systems in some detail. Before proceeding any further, however, we must refresh our acquaintance with the meaning and use of logarithms, which are necessary to us for both essential and technical reasons.

The role of logarithms can be best demonstrated to the musician by the example of octave relationships. Consider the respective string lengths of a series of descending octaves:

c	c_1	c_2	c_3	c_4	etc.
1	2	4	8	16	etc.

The ratios (1, 2, 4, 8, 16, etc.) form a series called a "geometric progression." In such a progression, the ratio between each two neighboring elements remains constant whereas the difference varies. Because the numbers 1, 2, 4, 8, and 16 are all powers of 2, we may write:

c	c_1	c_2	c_3	c_4	etc.
2^0	2^1	2^2	2^3	2^4	etc.

If we space these octaves according to string lengths, we obtain the following picture:

$$c \,. \,. \,c_1 \,. \,. \,. \,. \,c_2 \,. \,. \,. \,. \,. \,. \,. \,. \,c_3 \,. \,. \,. \,. \,. \,. \,. \,. \,. \,. \,. \,. \,. \,. \,. \,. \,c_4 \quad \text{etc.}$$

Figure 99

The octaves are not equidistant. But this is not the way we hear them. We do hear them as equidistant. They actually appear to us as they are spaced on a keyboard. In other words, we do not perceive them as equal ratios. Two octaves are twice as much as one; three octaves, three times as much as one; and so forth. The distances (1, 2, 3, 4, etc.) form a series called an "arithmetical progression." In such a progression, the difference between each two neighboring elements remains constant whereas the ratio varies.

Looking again at our series of descending octaves, where we expressed the ratios by powers of 2, we observe that the exponents of 2 represent an arithmetical progression. Now in an equation such as $8 = 2^3$, the exponent 3 is called the logarithm of 8 on the base 2. A logarithm is the power to which a number called "base" has to be raised in order to obtain a third number. Logarithms reduce the operation of multiplication to one of addition. Our hearing transforms the geometric progression of the string lengths into an arithmetical progression of octaves. We actually hear logarithmically, that is, we hear the logarithms of the string ratios (or of the frequencies).

It is relevant to consider in these terms the difference between harmonic and melodic hearing. Take the following example:

Figure 100

Harmonically, there is no difference between (a) and (b). The third remains a third whether it is octave-transposed or not. This is not at all true for the melodic relationship. Here the spatial position of the tones is of primary importance. Harmonically, we may disregard the tone space. Melodically, we may not. Clearly, all additions or subtractions of intervals are melodic, that is, they are spatial operations. Because the tone space has a logarithmic relationship to harmonic ratios, we must multiply ratios when we add intervals, and divide ratios when we subtract intervals. When we accumulate the same interval n times, we must raise its ratio to the nth power. Conversely, when we divide a given tone space into n equal parts, we must extract the nth root from the ratio of that tone space.

Here are a few examples, all in reference to $c = 1/1$:

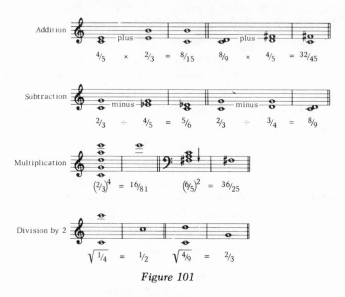

Figure 101

The examples of space division given above are all simple, for they involve only square roots that are rational. The task of dividing an octave, let us say, into 3 equal parts involves the cubic root of 2. Should we wish to divide the octave into 12 equal parts, we have to extract the 12th root of 2. In this latter case we want to find that number which, multiplied 12 successive times with any given frequency, gives twice that frequency—an operation that corresponds to the piling up of twelve equal intervals amounting to an octave. As we turn from the consideration of the essential role of logarithms in the relationship between the physical and the musical measures of pitch to practical applications of logarithmic reckoning, we must constantly keep in mind the deep-seated adequacy of such reckoning. As we have shown, logarithms are not only a mathematical convenience but they are intimately connected with our very apperception of pitch. This kind of connection is not unique, for logarithms are important in the understanding of a variety of phenomena, for instance, of organic growth. This insight inspired the Swiss mathematician Jakob Bernoulli (1654-1705) to refer to a certain logarithmic spiral as an "appearance of eternal truth that approaches secrets of faith." As a link between the psychic and physical worlds, logarithms served in the formulation of the Weber-Fechner law, which states that, in general, the response of any of our senses grows in intensity as the logarithm of the stimulus (cf. p. 61). Logarithms are currently used in music for measuring loudness. Admittedly, the Weber-Fechner law gives at best a rough approximation of the intensity of sensory responses. In the case of musical pitch, however, it is applicable with mathematical exactness.

When the concept of tonality began to draw ever more distant keys into its orb, the quest for purity of the triad gave way to the urge for modulation. The solution found toward the end of the seventeenth century is still with us (from which one may rightly conclude that the expansion of tonality is the unifying style characteristic of the last three hundred years). A finite octave system, as we have shown, cannot be built with pure triads on all twelve tones. If modulation is to operate unhampered, all intervals from any tone must therefore be made identical with all corresponding intervals from any other tone. There is only one answer to this postulate, and that is equal temperament. The octave is divided into 12 equal half-tones. The result is sacrifice of the purity of every interval except the octave, which remains pure. The natural beauty of the triad, too, is impaired; but modulations into all keys become equally possible, if equally out of tune.

In equal temperament, the size of each half-tone is the 12th root of 2, an irrational number. Accuracy becomes an issue. Because all tones tested on the sonometer within the octave lie on string lengths between 60 cm and 120 cm, we need not push accuracy beyond the millimeter. Even so, no less than six-place tables of logarithms are necessary for computation.

The 12th root of 2, which is the basis for all calculations in equal temperament, has the following logarithm:

$$\sqrt[12]{2} = \log 2 \div 12 = 0.30130 \div 12 = 0.02508.$$

Its antilogarithm, that is, the actual size of the 12th root of 2, is 1.0594. By multiplying 60 cm (the location of c^1 on the sonometer), or by dividing 120 cm (the location of c on the sonometer) successively by this number, we can determine the size and location of all tempered intervals:

Interval	Pitch	Ratio	String length (in cm)
Unison	c	1.0000	120
Minor second	$c\sharp$ ($d\flat$)	1.0594	113.3
Major second	d	1.1225	106.9
Minor third	$d\sharp$ ($e\flat$)	1.1892	100.9
Major third	e	1.2599	95.24
Fourth	f	1.3348	89.9
Augmented fourth	$f\sharp$ ($g\flat$)	1.4142	84.85
Fifth	g	1.4983	80.01
Minor sixth	$g\sharp$ ($a\flat$)	1.5874	75.6
Major sixth	a	1.6818	71.35
Minor seventh	$a\sharp$ ($b\flat$)	1.7818	67.35
Major seventh	b	1.8877	63.57
Octave	c^1	2.0000	60.00

The ratio column above represents the successive powers of $\sqrt[12]{2}$, by which the string length must be divided to determine the size of the respective ascending intervals, or multiplied to determine the size of the respective descending intervals. We calculate the tempered third e, for instance, by dividing the given c by four times the 12th root of 2 (four semitones).

For many purposes, it is practical to use a logarithmic system which is enlarged by the factor 1200/log 2. In this system, which was introduced by the Englishman Alexander J. Ellis in the nineteenth century, each half-tone is measured as having 100 cents. The octave, therefore, has 1200 cents. In this cent system, the figure for a given interval i is computed according to the formula 1200/log 2 $\times$ log i, where 1200/log 2 equals 3986. The transformation of cents into string lengths is more cumbersome by one step than that of ordinary logarithms. Therefore a graphic representation from which string lengths can be read off directly will facilitate the study of various divisions of the octave. Figure 102 shows how such a graph may be constructed. It should be designed to correspond to the actual string length of 120 cm. The graph indicates only the divisions corresponding to the 12-tone temperament, but the octave column to the left may be divided into any desired number of equal steps. The string lengths may then be read directly from the graph.

$$\text{Cents} = \frac{1200}{\log 2} \times \log i$$

$$\log i = \log m - \log n; \text{ e.g., } \log (c - b) = \log 24 - \log 13$$

$$\frac{1200}{\log 2} = 3986$$

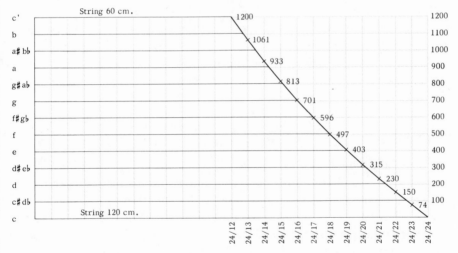

Figure 102

One need not suppose that the equal division of the octave results in equal deviations for all intervals. Equal temperament is basically an adjustment of twelve fifths to the octave. In other words, the Pythagorean comma is distributed equally over twelve fifths or seven octaves. The error for each fifth is one twelfth of the Pythagorean comma, which is much smaller, anyway, than the diesis (125/128) resulting from the superposition of three pure thirds.

One remembers from our calculation of b-sharp reached by fifths as compared to b-sharp reached by major thirds (cf. p. 217) that the former lies 0.81 cm from the octave; the latter (in the opposite direction), 1.44 cm. The fifths, in short, will be flat by a very small amount. On our sonometer, g (one fifth above $c = 120$ cm) lies at 80.1 cm in equal temperament as against 80.0 cm in pure intonation—a difference not noticeable, for all practical purposes.

The whole tone in equal temperament is necessarily flat by 2/12 of the Pythagorean comma. On the sonometer it lies at 106.9 cm as against the pure whole tone at 106.64 cm—again a negligible deviation of less than 3 mm.

The difference of two whole tones in equal temperament and pure intonation would be just as negligible, were it not for the intrusion of another norm at this point: the pure third as a constituent of the pure triad. In the triad, the Pythagorean third is precisely *not* the relevant norm. On the sonometer of 120 cm, the respective positions of the pure, Pythagorean, and tempered thirds are the following:

	Pure	Pythagorean	Tempered
Major third	96 cm	94.72 cm	95.2 cm

The musical meaning of these figures can be heard on the sonometer. As we sound the three tones one after the other, we perceive that the tempered third lies nearer to the Pythagorean third than to the pure third. Consequently, the mental adjustment in hearing the triad is rather large. The triad is the real victim of equal temperament. Actually, equal temperament results in a tempered diatonic Pythagorean scale, and not in a tempered just scale, for which the applicable criterion would be the purity of the triad.

The role of the fifth as the norm of equal temperament manifests itself also in the technique of tuning keyboard instruments. Tuning is done by ascending fifths and descending fourths. In order to meet the octave, each fifth is contracted by a very small amount, and each fourth (the inversion of the fifth) is reciprocally expanded. Thirds are used as checks as they become available. After the chromatic scale of one octave has been set, the rest of the keyboard is then tuned by pure octaves. Here is a schematic diagram of the process:

by tuning fork

Figure 103

But how does the tuner know by just what amount to contract the fifths, and to expand the fourths, so that he really will meet the octave after completing the circle of 12 fifths? Sometimes, of course, a tuner will at first land above or below the desired goal of the octave, and he will accordingly have to correct and repeat the process. As a guide to tempering, he calls to his aid the phenomenon we know as "beats" (cf. pp. 67 f.). To assume that he actually counts the beats when tempering the intonation of a keyboard is a popular illusion. He no more perceives beats as a quantity than he perceives a tone as a rate of frequency. What the tuner perceives is a certain *musical* change of character in respect to the pure fifth. The overtones of any tone, tempered or not, are, of course, pure in relation to the fundamental. Therefore the third partial of $a_1 = 220$, for instance, is $e^1 = 660$. This frequency would be identical with that of the second partial of e, the fifth of $a_1 = 220$, if this e were pure. But it is tempered; it has a frequency of only 659.25508. The difference between the two frequencies is 0.74492 per second, or 44.6952 per minute. This is the number of beats per minute between the pure and the tempered e^1. The tuner, then, is supposed to contract the fifth a - e^1 so that it will beat precisely 44.6952 times per minute. Because the number of beats always equals the difference between two absolute frequencies, the same fifth one octave higher will produce twice as many beats, or 89.3904 loudness fluctuations per minute. The tuner, under the circumstances, proceeds musically, not mechanically. This is the reason for the existence of good tuners and bad tuners. Long training and experience enable a good tuner to judge the deviation for the various fifths and fourths.

It is true that even the best tuning might not be completely accurate, that is, precisely equally tempered. It is also true that one could easily produce a set of twelve tuning forks for each tone of an equally tempered octave and then tune each tone in unison with the corresponding fork. Strangely enough, tuners have always resisted this kind of help; and it may be just as well. Ears go less easily out of order than do the best mechanical devices. As simple an instrument as a tuning fork is subject to temperature changes; and twelve forks, of different sizes, will vary unevenly. Above all, the thought is rather comforting that tuning is still dependent upon personal sensitivity and skill rather than upon mechanical appliances and devices.

Equal temperament, although the only answer to the style of music that has developed since Bach, is admittedly a somewhat brutal operation. It has brought about a situation in which the actual connection with some norms has grown extremely tenuous. Often the correct orthography can be discovered only by reading. Some people claim that the increasing importance of the written pitch symbols indicates a trend toward spiritualization. On the other hand, the danger of rupturing the connection with norms and consequently of falling into a new materialism is demonstrated by the theory of twelve-tone music. Here, all interpretation of tones in the direction of norms has been abandoned, and the division of the octave into twelve equal spaces is taken at face value, as a simple *fait accompli*, rather than as a necessary compromise. Seen in this light, the long resistance against universal acceptance of equal temperament should be considered with due respect for those musicians who hesitated to abandon their foothold—who insisted on actually hearing what they were supposed to hear, in preference to interpreting a false sound as something else they thought it should be.

In the centuries before the introduction of equal temperament in Europe, all systems of tempering were directed toward the preservation of as many pure triads as possible. Rather than distributing the error equally among the 12 tones, as is done in equal temperament where all triads are false but only minimally so, organ tuners in the early seventeenth century, for instance, rescued the purity of some basic triads by letting a more removed triad bear the whole brunt. To be sure, that crucial spot (e.g., one involving a choice between G-sharp and A-flat) was so much out of tune that it acquired the name of "wolf." Systems of this kind were all "unequal" temperaments, in which the inescapable false triad was pushed back to here or there, according to personal or stylistic preferences. Of all these systems, only one reached a position of practical importance, as we remarked earlier, namely, the meantone temperament. It existed in a number of variants, but the main principle emerges clearly. The major third E is tuned to the pure "5" (i.e., it coincides with 1/5 of the string length or with the 5th partial), and the fifths are contracted so that four fifths reach 80 instead of the Pythagorean 81. In a way, meantone temperament is a reverse of Pythagorean tuning where the fifths stayed pure and the thirds were expanded.

In the process of tuning a keyboard according to meantone temperament, the first conflict between the upward and downward projections of the contracted fifths occurs three keys away from the center. In relation to C major, the tuner has to decide between G-sharp in A major on one hand, or A-flat in E-flat major on the other; for G-sharp chosen as the pure third of E is not identical with A-flat obtained by a downward projection of the contracted fifths. C-sharp and F-sharp are usually tuned as pure thirds to A and D, respectively. B is gained either as a pure third to G or as an extension of the row of ascending fifths.

In the computation of the tones forming the meantone temperament, the basic problem is the division of the tone space c_2 - e into four equal fifths:

Figure 104

Because the third in meantone temperament corresponds exactly to 5, the value for each of the four fifths is the fourth root of 5, or 1.495.[1] Accordingly, each downward fifth (e.g., f_1 in relation to c) corresponds to $\sqrt[4]{5}$, and each upward fifth to $1/\sqrt[4]{5}$. Successive fifths figure as powers of these two expressions. We find the upper major third of any of these tones (e.g., f-sharp, g-sharp, etc.) by dividing the ratio of the appropriate fifth by 5. The following table shows a scale tuned in meantone temperament, with the corresponding string lengths for the octave lying on the sonometer between 120 cm and 60 cm:

---

[1] Here is the logarithmic computation for those who do not wish to take this number on faith:
$$\frac{\log 5}{4} = \frac{0.698970}{4} = 0.1747425$$
$$\text{Antilog } 0.1747425 = 1.495$$

Interval	Pitch	Ratio	String length
Unison	c	$1/1$	120.0 cm
Semitone	$c\sharp$	$16/5(\sqrt[4]{5})^3$	114.9 cm
Major second	d	$2/(\sqrt[4]{5})^2$	107.3 cm
Minor third	$e\flat$	$(\sqrt[4]{5})^3/4$	100.3 cm
Major third	e	$4/5$	96.0 cm
Fourth	f	$\sqrt[4]{5}/2$	89.7 cm
Augmented fourth	$f\sharp$	$8/5(\sqrt[4]{5})^2$	85.9 cm
Fifth	g	$1/\sqrt[4]{5}$	80.2 cm
Augmented fifth	$g\sharp$	$(4/5)^2$	76.8 cm
Minor sixth	$a\flat$	$5/8$	75.0 cm
Major sixth	a	$2/(\sqrt[4]{5})^3$	71.8 cm
Minor seventh	$b\flat$	$(\sqrt[4]{5})^2/4$	67.1 cm
Major seventh	b	$4/(\sqrt[4]{5})^5$	64.2 cm
Octave	c^1	$1/2$	60.0 cm

It is imperative to adjust the sonometer to the given string lengths and to listen attentively to the meantone scale as well as to the various triads. One might be surprised to find the wolves less "howling" than expected, at least on a stringed instrument. On an organ, the false tuning is more disturbing. There is good reason to believe that composers writing for an unevenly tempered keyboard avoided critical triads by renouncing certain keys. An occasional occurrence of false chords could be glossed over, for instance, by the introduction of fluid ornaments. A study of ornamentation in keyboard works of the Baroque might well proceed from acoustical considerations prompted by meantone temperament. As to our own reaction, the prolonged contact with equal temperament has probably made us less sensitive to deviations than musicians in earlier centuries may have been whose ears were still attuned to pure harmonies. How many are there among us who have ever heard the sound of a pure triad?

In concluding, a word about microtones may be in order, that is, about intervals that are smaller than the half-tone. The half-tone happens to be the smallest interval in traditional Western music, but the discovery of smaller norms is entirely possible. Anyone probing into the microtonic world must keep a few guiding principles in mind. First, micro-intervals result from a deeper penetration into the structure of tone and will therefore not eventuate in a discarding of previously discovered norms. Second, just as we have not reached the half-tone by dividing the octave, so we shall not find any new primary interval by splitting the half-tone. The new building stones must be created by the discovery of new harmonic norms and not through a division of the octave by some number greater than 12. Third, any new scale thus found will eventually have to be tempered. Tempered intervals can never be primary norms, for they are the result of a man-made compromise. But once a harmonic norm is established, then the octave can be divided by the appropriate figure for the purpose of fitting the norm into the octave; and a new tempered scale is produced. Attempts to introduce micro-intervals have been successful whenever the approach was harmonic and not mechanical, for example, in the enharmonic scale of the Greeks, the *sruti* of the Hindu system, and, to some extent, the Renaissance experiments of Nicola Vicentino. They have failed whenever norms have been erroneously ignored, for example, by the forceful Alois Hába (b.1893) and by most electronic composers.

Exercises

1. Realize an equally tempered scale on the sonometer.
2. Realize an unequally tempered scale on the sonometer (e.g., the meantone scale as described in this chapter).
3. Calculate, and realize on the sonometer, a meantone temperament with A-flat instead of G-sharp.
4. Calculate, and realize on the sonometer, a meantone temperament in which *B* is tuned as the pure third of *G*.

15 | Conclusion: The Realm of Acoustics

Acoustics, involving a scientific endeavor and a quest for knowledge, necessarily follows the trends and fortunes of scientific thought in general. In the acoustics of music, however, an element is present that may be disregarded in other sciences, namely, a preoccupation with value. In musical acoustics we do indeed concern ourselves with a search, not purely into the nature of anything, but only into the nature of such things that constitute the basis of music. However widely or narrowly we may define artistic freedom, and however large or small a part we may ascribe to that which is binding (be it thought of as residing in physical, physiological, psychological, or metaphysical principles), we can never completely disregard, and indeed never are disregarding, value. Acoustics of music always carries an axiological element, whether we are aware of it or not; and the history of the acoustics of music may be looked upon as a quest for knowledge surrounding the question: why are some or all of the fundamental musical facts what they are?

In the course of time, the gaze of man seeking knowledge and guidance has shifted from the beyond to the outer world, and from the outer world has turned to the inner world. Accordingly, science was first metaphysical, then physical, and later psychological. This order does not exclude the coexistence of the several stages but rather points to a general direction of interest at a given time. We do not forget that an Aristotle is always to be found beside a Plato, and a Roger Bacon beside a St. Thomas Aquinas. Nor do we wish to associate the concept of unidirectional development with the idea of progress. We do not presume to draw any conclusion as to the form which the succession of stages might eventually assume—straightforward, or periodic, or more complex.

The question of method need not detain us here. We merely play with shifting accents when we call deduction suitable to metaphysics, and induction suitable to modern science. One method really never exists without the other. The acoustics of antiquity and of the Middle Ages was ruled by Pythagoras. The shift from the Greco-Medieval to the "new" acoustics was slow and gradual, as it was in other scientific fields and in science at large. The landmarks along the road are great and representative names, such as Gioseffo Zarlino (1517-1590), Marin Mersenne (1588-1648), and Joseph Sauveur (1653-1716). Within a generation of Sauveur, who established the overtone series, Giuseppe Tartini (1692-1770) and Georg Andreas Sorge (1703-1778) discovered the difference tones. Jean-Philippe Rameau (1683-1764) marks a turning point: his first work, *Traité de l'harmonie* (1722) is still essentially arithmetical, but the *Démonstration du principe de l'harmonie* (1750) is already based on the natural phenomenon of the overtone series. Hermann von Helmholtz (1821-1894) marks another point of change. His *Lehre von den Tonempfindungen als physiologische Grundlage für die Theorie der Musik* (1863) isolates and describes the essentials of what happens "out there." It freed the way to the tracking of tone within the subject; and subsequent investigations shifted to physiology, psycho-physiology, and psychology. Carl Stumpf (1848-1936) is generally regarded as the initiator of the psychology of music. The foremost specialist of the ear among contemporary physiologists is probably Georg Von Békésy, who was awarded the Nobel Prize for medicine in 1961.

The musician, true to his axiological viewpoint, will inquire into the meaning of all these attitudes and approaches. There are many layers to the question, from which we shall peel off only a few. What is the effect of scientific investigation upon music? What is the influence of music on scientific investigation? How are the scientific findings to be interpreted?

Restricting ourselves for the moment to the "new," scientific acoustics, we can observe only one instance of an apparent, although not undisputed, influence of an acoustical discovery on music. The age of the Enlightenment was most anxious to justify everything artistic by nature. Rameau's delight with the natural overtone series was intense. Because the series contains the major but not the minor triad, the domination of the latter by the former seemed proven beyond all doubt. "That first burst of nature is so powerful, so brilliant, so virile—if I may call it thus—that it surpasses minor and shows itself to be the master of harmony," exclaims Rameau. The scientific discovery of the major triad in the overtone series may well have been at least partly responsible for the exalted position of major over minor from the eighteenth century until practically our own time.

We approach the second question—concerning the influence of music on scientific investigation—by quoting the eminent physicist Henri Bouasse: "Is it necessary to add that excellent music was composed before the scientific question had come up? Fortunately, musicians may show genius without knowing the first thing about acoustics, and the physicist entertains no hope that his theories will provide them with even the slightest inspiration. In turn, whether his theories win the approval of the musicians is to him of no concern."[1] This statement sounds like a defense rather than a renunciation. For indeed, what is the object of general acoustics? The audible. What is the object of musical acoustics? The audible, restricted to the musically usable, such as tone, intervals, chords. Suppose an acoustician would come up one day with a musically absurd conclusion, such as the perfect dissonance of the octave, and suppose he would actually perceive the octave thus through some malfunction of his hearing apparatus. Would he dare maintain his position in the face of the overwhelming musical evidence? Hardly so. The function of theories is to explain phenomena. The function of musical acoustics is to explain musical phenomena. Negating this relationship is like pulling the bread from under the knife that is supposed to cut it.

The lack of reciprocal influence between the science of acoustics and the art of music is of relatively recent date, not to be observed in the "old," classical acoustics. It springs, largely speaking, from the fragmentation of human pursuits since the Renaissance and, more specifically, from the concomitant independence and professed objectivity of science. Objectivity means truth in the sense of adequacy to the object. Only in this sense can it be said that science must pursue its goals with complete independence. In the natural sciences, the behavior and the laws of nature are the truth to be pursued. In musical acoustics, the norms and the laws of musical man are the truth to be pursued. No other kind of objectivity can be valid.

Our third question, which concerns the interpretation of scientific findings,

---

[1] *Bases physiques de la musique* (Paris, 1906), p. 76. Transl. by the authors.

is well suited to serve as a transition to a characterization of the "old," Pythagorean acoustics.

Let us proceed from a firmly established, undisputed fact. We submit, as an example, our sense for symmetry. At this point, we prefer an example outside acoustics precisely because, in the minds of most people, musical facts are infirmly established and difficult to interpret. This dialectic caution need not prevent the reader from thinking of the triad, let us say, while discussing symmetry.

Symmetry is a clear norm to us. This truth is mirrored even in language. There is no positive word describing the contrary of symmetry. Instead, we must use such negative expressions as "lack of symmetry," "asymmetrical," "departure from symmetry," and the like. Symmetry, then, is an inborn value concept. Up to this point, there is likely to be general agreement. But what happens when we initiate a discussion on the significance of the observed fact? Somebody will venture the remark that our body is visibly symmetrical. Another person will add that this symmetry of our body makes the quality of symmetry in general appear normal and valuable to us. A majority—but no longer a unanimous group—will agree that the structure of our body determines that of our mind. Against this opinion, a few will argue—and we wish to include ourselves—that one could perhaps conceive of a general principle of symmetry that manifests itself in both body and mind. The majority shudders at the heresy and accuses the few of mysticism—a veritable insult when hurled at a contemporary mind. We cannot, however, feel either insulted or dismayed, for indeed we detect neither more nor less mysticism in our opinion than in that held by the majority. First principles are in fact always mysterious, and it is the fate of value problems to be located on the threshold of the unknown. If body symmetry is the cause of the general value of symmetry, we might, for instance, ask, what causes body symmetry? And if the answer should suggest that body symmetry is necessary for mechanical reasons and that an asymmetrical body is impractical or unfit for life, then clearly the concept of teleology has crept in—a mystical idea if there ever was one.

The nature of first principles precludes any experimental proof. They may be statistically outlined, but the gathering of statistical evidence is insufficient to approach the character of proof in any scientific sense. One must therefore realize the futility of discussing first principles beyond the point of agreement or disagreement.

This realization does not absolve anybody from having to decide for himself which opinion to adopt. Especially in a field like musical acoustics, such a decision is absolutely necessary; for without the light emanating from principles, the facts remain shapeless and meaningless. Two guiding precepts may be offered to the struggling mind. For once, in a limited field like that of musical acoustics, statistical evidence need not be denied *some* value, in the sense that a principle—considered at least as a working hypothesis—will show its worth with increasing fecundity of application. The works of the master composers—a vast and long laboratory experiment—teach us a great deal and prove the principles by recurrent and cumulative demonstration. Second, the beauty or the elegance of a concept is an intrinsic value in itself, as mathematicians well know. The authors of this book have let themselves be guided by both these criteria. The resulting viewpoint turns out to be within the Pythagorean tradition. What is the Pythagorean tradition?

Pythagoras is one of the few historic figures who have attained legendary grandeur, not by writing, but solely by teaching. He flourished in the sixth century B. C. which also saw Confucius, Lao-tse, Buddha, and Zarathustra live and die. His life is known in dim outlines only. Of his work, every schoolchild knows the theorem that bears his name; and every student of music, his experiments with a vibrating string. Our knowledge of his philosophy, however, is beclouded by the protective secrecy with which it was deliberately surrounded by the Pythagoreans. Cleared of its exoteric encumbrances, the core of the Pythagorean acoustical philosophy emerges in the following description.

The tone phenomenon presents two aspects: an outer one and an inner one. "Out there" is the vibrating string with its attributes—all measurable, quantitative, material, haptical. "In here" is the tone proper—nonmeasurable, qualitative, nonmaterial, musical. These two sides of the acoustical happening are essentially different. They are, in fact, incomparable, but they can be linked by number. The specifically Pythagorean evaluation takes place at this point. Whereas we might be tempted to declare: "After all, tone is nothing but a vibration," Pythagoras would rather have said: "After all, vibration is nothing but a tone." This dialogue may be a rhetorical exaggeration but serves to accentuate a difference in the trend of interpretation. Whereas modern man is likely to derive most satisfaction from reducing qualities to quantities, the enthusiasm of Pythagoras was, on the contrary, aroused by the fact that mathematics could be heard. He took the inner phenomenon of the musical tone at least as seriously as its outer mechanical counterpart. Once this position is established, the next step follows necessarily. The linking element, namely, the number representing both a haptical ratio and a musical interval, may now be charged not only with an arithmetical but also with a qualitative meaning. Number will indeed rise from a mere figure to a symbol for quality as well as for quantity. Such a number is itself a *morphé* and possesses shaping power. Number is thus taken to be the shaping force behind both the string ratio and its inner counterpart, the tone. It is therefore a first principle; and in this sense the Pythagoreans could say, "All things are number." Modern science can ill afford not to subscribe to this statement but has done so only in respect to the haptical side of number.

This, in simple language, is the basic doctrine of Pythagoras. Upon the metaphysical (but not mystical) principle of number, the followers of Pythagoras conferred the highest possible dignity of a cosmic first principle. Modestly, the acoustician might be satisfied with a more restricted application. Confined to the domain of its origin, musical acoustics, the Pythagorean principle elucidates many points of theory while frequently opening up more distant perspectives.

In the opening paragraph of this chapter, we have alluded to the question of artistic freedom. In these closing paragraphs, we may properly revert to it. The statement has often been made—and less often understood—that freedom without law is meaningless. In the social context, this word—we may confess it—easily takes on a suspicious tinge, for political states speaking on their own behalf have been known to emit comparable maxims unworthy of our respect. Yet there never was a statement more correct or more universally true. The whole phenomenal world bears witness to it. The uncreated alone is limitless. The process of phenomenalization is one of shaping; and shaping means law, limitation, restriction, definiteness, discreteness, individuation. The process is paralleled, on its proper level, in artistic creation. The stone of the sculptor, the canvas and the pigments of the painter, the discrete tones of the musician: they all are limiting elements, necessary to art because distinguishing it from nature. Although the line of separation is sometimes as thin as a hair, the domain of art extends no further than to the borders beyond which absolute realism lies. The prime condition of art is its distinctiveness from nature.

From the beginning, music is removed from the pitfall of becoming a simulacrum of nature, which in the other arts has occasionally claimed a victim. This observation may at once be qualified by another one, namely, that music, compared to the other arts, symbolizes a deeper, more essential layer in nature. The existence of musical acoustics is significant in this context. No discipline in the visual arts can be said to correspond to musical acoustics—architecture excepted, which is by nature more akin to music than to painting or sculpture. Whatever in optics could relevantly be called "pictorial optics" or "sculptural optics" is surely a long way removed from the immediate and deep-seated relationship existing between acoustics and music.

It is because of the remoteness of music from the phenomenal world that acoustical data assume such a peculiar character. So fundamental and so intimate are they that a differentiation between technical and normative factors appears difficult. Indeed, what has sometimes been called the "raw material" of music—and in a certain sense correctly so—is quite removed from the state which we commonly associate with rawness. Not only is even the single tone already a refined product extracted from noise, but it is in addition structured in an actual and in a potential sense. Tone is the seed containing the normative morphologies of melody and harmony.

When Leibniz said that music is "a secret mathematical exercise of the soul," he was stressing that intimacy of the connection between acoustics and the "art of acoustics" : music. But he was also performing a reduction to quantity, which, as we now know, leads to a dead end. This partiality was recognized by Schopenhauer who called Leibniz' definition true in a "lower" sense and then corrected it to read: "Music is a secret metaphysical exercise of the philosophizing soul." He knew that "the phenomenal world (nature) and music are two different expressions of the same thing." Beethoven went so far as to assert most emphatically: "Music is a higher revelation than all wisdom and philosophy." In another context, Pascal equated the qualitative and measurable sides by saying: "They know by the heart as others know by the mind." For Bach, counterpoint was a proof of the existence of God.

These few quotations, which could easily be increased to the volume of a whole book, bear witness to an awareness—more widespread than one might perhaps at first suppose—of the real meaning of music. Indeed, only by recognizing the epistemological equivalence, that is, the equal value, of tone and number, we can do justice to the total phenomenon, which manifests itself in the inner as well as in the outer world.

Appendix A

Divisions of the monochord string (as explicated on pp. 14 through 35) can all be exactly performed by the ear alone without any help from a measuring tape.

Division by 2 calls for common sense. If the two halves are to be equal in length, they will also have to be equal in sound. The eye can find the approximate middle of the string. The ear will identify the exact middle by listening for the same pitch on either side of the dividing point. Either of the two halves, which are in unison to each other, sounds against the control tone of the whole string the musical interval of the octave.

We now use an octave to divide the string by 3. The point at 1/3 of the string separates two sections, 1/3:2/3 = 1:2, which sound an octave against each other. We recognize the new interval sounded by the longer section against the control string as a fifth, 2/3:3/3 = 2:3. The shorter section, one octave higher, sounds the twelfth, 1/3:3/3 = 1:3.

Division by 4 and by all subsequent even numbers can be performed in two different manners (which may be used to verify each other). By one method, one proceeds from the string length to be divided by 2, testing for unisons. Thus to divide by 4, one first pinpoints 1/2 of the string and then takes 1/2 of that length in the already practiced manner.

By the other method, one always applies the gain of one division to the next one, listening to the interval between the two string sections. Division by 4 produces the ratio 1/4:3/4 = 1:3, the musical twelfth familiar from the preceding division by 3. Division by 5 produces the ratio 1/5:4/5 = 1:4, or the double octave. Against the control tone of the whole string, the remainder 4/5 yields the musical interval of the major third. This new interval now makes aural division by 6 possible; for when the two string sections give us the musical experience of a major third (two octaves apart), the lengths of the two sections relate to each other as 1/5:5/6 = 1:5.

241

This procedure may be continued until the ever bigger disparity between the lengths of the two string sections puts a practical halt to the aural adventure. How to reach as complicated a ratio as the Pythagorean comma, six digits over six digits, without the help of a measuring tape, the following Appendix B demonstrates.

Appendix B*

by Ernest G. McClain

Tuning theory is best demonstrated on the monochord, an instrument designed to permit the sounding length to be varied by a movable bridge without altering string tension, and of sufficient length to make commas and other micro-intervals both visible and audible.

The arithmetical complexities involved in defining the various tuning systems and the commas between them can be avoided by a simple paper-folding exercise analogous to the "rope-stretching" by which land was once measured and the proportions of buildings determined. The folding is done most conveniently on a strip of paper one or two inches wide and three to four feet long; adding-machine tape serves very well. The examples described here were developed on a string and paper length of 120 centimeters. The pitches located at successive creases should be sounded on the string to reinforce the lesson for eye and ear.

The following exercise is carried through four stages.

(1)   Eight successive folding operations establish the diatonic major scale in Pythagorean tuning.

(2)   Five more folds establish the chromatic semitones needed for modal permutations, and a sixth fold reaches the Pythagorean comma 531441:524288.

(3)   Two folds establish the pure thirds of 5:4 needed for the Greek Dorian scale in Just tuning (related to the *Republic* scale), involving syntonic commas of 81:80; and two more establish the diesis 128:125.

(4)   The syntonic commas 81:80 are split visually to determine the approximate location of several tones in equal temperament.

Important insights are gained at each of the above four stages; failure to complete the whole series does not diminish satisfaction earned along the way. If the folding is done carefully, the errors involved will prove subliminal when tested on the string.

* Reprinted, with the kind permission of the publisher, from Ernest G. McClain, *The Pythagorean Plato: Prelude to the Song Itself* (Stony Brook, New York: Nicolas Hays, 1978), 169-75.

Exercise 1: The C Major Scale in Pythagorean Tuning

Cut the strip of paper exactly the same length as the sounding portion of the string. Mark the left end 0 (meaning zero string length) as a reference point for all folding operations, and mark the right end $\frac{1}{1}$ = B as the schematic ground tone sounded by the full length of string. Later we shall dispense with this lowest tone, treating it as the Greeks treated the ground tone on their monochord, calling it *Proslambanomenos* ("added tone") and leaving it outside their tetrachord system. If folding always proceeds from 0 on the left, it will produce the descending octave scale on the right half of the paper strip. (If 0 were placed on the right and folding started from there, the same scale would appear ascending from the left.)

The first folding operation is the most important: it establishes the model for all that follows. Bring the two ends of the paper together and crease, then bring the doubled ends together and crease again. You have now divided the string into four parts: label them $\frac{1}{4}$ = B″, $\frac{1}{2}$ = B′, $\frac{3}{4}$ = E, making a vertical mark along each crease, whose locus will otherwise soon fade.* Test on the string by sounding the successive lengths. Note the descending perfect fifth 2:3 at B′:E and the complementary perfect fourth 3:4 at E:B. Future folds are planned to project all tones into this octave.

Now consider these first results carefully to glean the two principles used from hereon:

(1)   To ascend a perfect fourth from any tone, equivalent to a multiplication by $\frac{3}{4}$, merely quarter its sounding length by folding in half twice (folding always from 0 towards the tone) and locating the new tone on the third crease (as the new tone E appeared here). The new tone appears on the left.

(2)   Alternatively, to descend a perfect fifth from any tone, equivalent to a multiplication by $\frac{2}{3}$, fold the reference length in half (folding always from 0 toward the tone length), and then use this folded portion to measure a third equal segment along the remainder. The new tone appears on the right, as E appears to the right of B′.

These two folding principles produce alternate higher fourths 3:4 and

---

* Ernest McClain uses capital letters to designate pitches reached by fifths, and lower-case letters to designate pitches reached by thirds.

lower fifths 3:2 so that, like the piano tuner "laying the bearings" in the central octave, we keep all new tones within the compass of our chosen octave.

Now complete the heptatonic (7-tone) series by five more folds of the kind described above:

¾ E = A *(reminder: divide 0-E into fourths)*
½ A = D *(divide 0-A into halves, then measure along remainder)*
¾ D = G
½ G = C
¾ C = F

The seven tones from B to B′ actually define the modern Locrian mode in Pythagorean tuning. Now halve C to produce C′ and fold the end segment C:B back out of sight; the result is our familiar C major scale. Similar halving of successively higher tones would produce the Greek Phrygian octave D-D′, the Greek Dorian octave E-E′, etc., but they teach us nothing new.

In Pythagorean tuning all major seconds are the same size, 8:9, slightly larger than the wholetones of equal temperament. The two minor seconds at E:F and B:C are undersized semitones of 243:256, which the Greeks called a *leimma* ("left-over," as the difference between two wholetones and a fourth 4:3). This is the tuning which modern string players approximate when they "stretch" intervals to intensify upward or downward "leading-tones." Pythagorean tuning remains relevant to modern harmony in the sense that the most powerful line of relationships extends through the dominantic order of perfect fourths and fifths. Such intervals are the strongest anchors for a mind seeking "shape" within the tonal flux.

Exercise 2: The Chromatic Scale and Pythagorean Comma

A continuation of the procedure outlined above will produce an endless number of alternating fourths and fifths. Five more folds will produce the five new tones needed for a twelve-tone chromatic scale in Pythagorean tuning.

¾ $F = B\flat$ *(reminder: divide 0-F into fourths)*

½ $B\flat = E\flat$ *(divide 0-B♭ into halves, then measure along remainder)*

¾ $E\flat = A\flat$

½ $A\flat = D\flat$

¾ $D\flat = G\flat$

The chromatic scale shows alternate undersized diatonic semitone leimmas 243:256 (at C:B, B♭:A, A♭:G, etc.) and oversized chromatic semitone apotomes 2048:2187 (at B:B♭, A:A♭, G:G♭, etc.). One more folding operation produces a 13th tone (= 12th "disciple") which disagrees with the reference B' = ½ by a Pythagorean comma 531441:524288, an interval too small to allow the tones an independent status, yet large enough to produce an offensive disagreement which requires that one of them be eliminated from the set:

¾ $G\flat = C\flat$, *and C♭:B' is the Pythagorean comma.*

A continuation of our tuning procedure would simply produce further commas with each of the original 12 tones in turn, hence we have reached the tonal boundary of Pythagorean tuning. (Note: If we allowed ourselves to make a more awkward triple fold, ⟶, of the original reference length B = ⅓, we could locate at ⅔ = F♯ another Pythagorean comma G♭:F♯ Our procedure is designed to avoid these awkward, and perhaps logically inadmissible, triple folds.)

An appropriate selection from among the 12 tones produces the pattern of the ancient Greek Dorian mode, Plato's "true Hellenic mode," the pattern used in *Timaeus* for the World-Soul (in the C octave, not the D octave used in the text):

C		B♭		A♭		G		F		E♭		D♭		C
	t		t		s		t		t		t		s	

We notice here that the sequence of wholetones (t) and semitones (s) is exactly opposite to that of the C major scale.

Exercise 3: The Greek Dorian Scale in Just Tuning

By retuning two of the above tones we can produce the Greek Dorian scale in the so-called Didymus tuning, one of several forms of Just tuning, which always involves some mixture of fifths 2:3 and fourths 3:4 with pure thirds of 4:5. The folding procedure is a slight variant of that employed previously: quarter the reference length by a double fold, then measure a fifth segment at ¾ along the remainder.

¾ C' = a♭ *(starting always from 0, quarter the length for C', to determine the measure for a♭, to its right).*

¾ G = e♭ *(quarter the length for G, then measure e♭).*

The micro-intervals a♭:A♭ and e♭:E♭ are syntonic commas 80:81. (They are the difference between a pure third ¾ = $^{80}/_{60}$ and a ditone third $(9/8)^2 = {}^{80}/_{64}$.) The following tones and ratios are the Didymus tuning of the ancient Dorian octave.

C	B♭	a♭	G	F	E♭	d♭	C
8:9	9:10	15:16	8:9	8:9	9:10	15:16	

The pure triad harmonies which became important to Western music in the sixteenth and seventeenth centuries required another variant of Just tuning, involving similarly warring commas in the tone field. Musicians were forced to disguise them by one subterfuge or another. What could not be disguised, however, were the worse disagreements which arose when two or three pure thirds 5:4 were taken in succession (as, for instance, in nineteenth-century Romantic harmonic modulations). Two more folding operations by 5:4 expose the diesis 125:128 which lies in wait; it is the discrepancy between three pure thirds 5:4 and the octave 2:1.

¾ a♭ = f♭

¾ f♭ = d♭♭

The ratio d♭♭:C is $\dfrac{(5/4)^3}{2}$, or 125:128, the diesis.

Exercise 4: Equal Temperament

Paper-folding cleverness cannot demonstrate the idea of a number like the "twelfth root of 2" (approximately 1.059463+) which provides the theoretical basis for an equal tempered scale. But because equal temperament tones lie *within* the syntonic commas we have already established, we can estimate the location of several of them with almost as much accuracy as the ear can appreciate. Interpolate the tempered A♭ within the syntonic comma a♭:A♭ of Just Pythagorean tuning, the tempered E♭ within the syntonic comma e♭:E♭, and the tempered D♭ within the syntonic comma d♭:D♭, as illustrated in Fig. 105.

The ratio for an equal tempered semitone is one which carries us to the octave 2:1 after 12 identical operations. Vincenzo Galilei, famous father of an even more famous son, first explained how to achieve this to a sixteenth-century lute-maker, relying on the ratio 18:17—an approximation at which he arrived not by calculation, we are told, but by intuition. Like Vincenzo we trust our intuition that the equal tempered tones we located in the middle of syntonic commas are accurate enough for present purposes—that is, accurate enough to make clear that temperament is a compromise between the conflicting claims of various ideal values.

In tuning a scale we are haunted by the perfection which lies just beyond our grasp, and intrigued by the necessity for tempering even as much as lies within it. Paper-folding demonstrates the ear's problem to the eye: cyclic repetition at the octave requires that all smaller intervals be robbed of some measure of their perfection. None can be given "exactly what they are owed" if the system as a whole is to function at its best.

Location of Initial Tones

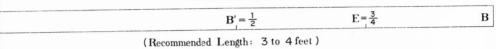

$B' = \frac{1}{2}$ $E = \frac{3}{4}$ B

(Recommended Length: 3 to 4 feet)

Sectional Views at Completion of Demonstration

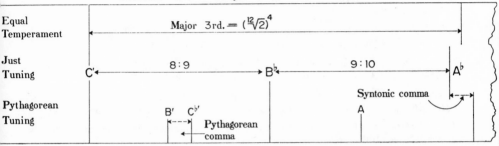

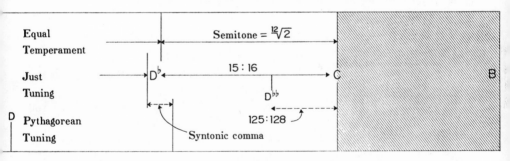

Figure 105. *Pythagorean Tuning, Just Tuning, and Equal Temperament on the Monochord*

SUGGESTED READINGS

Barbour, J. Murray. *Tuning and Temperament*. East Lansing: Michigan State College Press, 1953.

Bartholomew, Wilmer T. *Acoustics of Music*. New York: Prentice-Hall, 1942.

Benade, Arthur H. *Horns, Strings, and Harmony*. Garden City, New York: Doubleday & Company, 1960.

Berlioz, Hector. *Treatise on Instrumentation*. Enlarged and revised by Richard Strauss. Translated by Theodore Front. New York: E. F. Kalmus, 1948.

Bibliography of Hearing. Complied by Harvard University, Psycho-Acoustic Laboratory. Cambridge, Massachusetts: Harvard University Press, 1955.

Boehm, Theobald. *The Flute and Flute-Playing*. 2nd English edition, revised and enlarged, translated and annotated by Dayton C. Miller. New York: McGinnis & Marx, 1960.

Helmholtz, Hermann L. F. *On the Sensations of Tone*. Translated by Alexander J. Ellis. New York: Dover Publications, 1954.

Jeans, James. *Science and Music*. Cambridge: University Press, 1947.

Kayser, Hans. *Lehrbuch der Harmonik*. Zürich: Occident Verlag, 1950.

Knudsen, Vern O. *Architectural Acoustics*. New York: John Wiley & Sons, 1932.

Levarie, Siegmund. "Noise," *Critical Inquiry* IV/1 (Autumn, 1977), 21-31.

Levy, Ernst. "Goethes musiktheoretische Anschauungen," *Schweizerische Musikzeitung* XCII/10 (October, 1952), 7-15.

McClain, Ernest G. *The Pythagorean Plato: Prelude to the Song Itself*. Stony Brook, New York: Nicolas Hays, 1978.

Miller, Dayton C. *The Science of Musical Sounds*. New York: The Macmillan Company, 1916.

Rameau, Jean-Philippe. *Traité de l'harmonie*. Facsimile of the first edition (Paris, 1722). New York: Broude Brothers, 1965.

Richardson, Edward G. *Sound*. London: E. Arnold & Co., 1953.

———. *The Acoustics of Orchestral Instruments and of the Organ*. London: E. Arnold & Co., 1929.

Rose, Arnold. *The Singer and the Voice*. London: Faber and Faber, 1962.

Schlesinger, Kathleen. *The Greek Aulos*. London: Methuen & Co., 1939.

Vitruvius. *The Ten Books on Architecture*. Translated by Morris Hicky Morgan. New York: Dover Publications, 1960.

Von Békésy, Georg. *Experiments in Hearing*. New York: McGraw-Hill, 1960.

Wood, Alexander. *The Physics of Music*. 5th ed. London: Methuen & Co., 1950.

Index

251